T0259450

Pressure Vessel and Stacks Field Repair Manual

Pressure Vessel
and Stacks
Field Repair Manual

A. Keith Escoe

AMSTERDAM • BOSTON • HEIDELBERG • LONDON
NEW YORK • OXFORD • PARIS • SAN DIEGO
SAN FRANCISCO • SINGAPORE • SYDNEY • TOKYO
Butterworth-Heinemann is an imprint of Elsevier

Butterworth-Heinemann is an imprint of Elsevier
30 Corporate Drive, Suite 400, Burlington, MA 01803, USA
Linacre House, Jordan Hill, Oxford OX2 8DP, UK

Library of Congress Cataloging-in-Publication Data
Application submitted.

British Library Cataloguing-in-Publication Data
A catalogue record for this book is available from the British Library.

ISBN: 978-0-7506-8766-9

For information on all Butterworth-Heinemann publications,
visit our Web site at: http://www.books.elsevier.com

Typeset by Charon Tec Ltd., A Macmillan Company (www.macmillansolutions.com).

Printed and bound in the United Kingdom
Transferred to Digital Printing, 2010

To three special people: my nephew, Commander Mark A. Escoe,
Supply Corps, United States Navy; my niece, Barbie Davis;
and to the memory of their Dad and my brother, Creighton L. Escoe,
who died tragically in a fire on March 7, 2008.

Contents

About the Author   ix

Preface   xi

1. **Systems of Units**   1
   - Getting Familiar with Metric SI Units   4
   - Other Important Metric SI Units Used in Mechanics   5
   - Units for Stress and Pressure   6
   - A Warning About Combining Metric SI Units   6
   - Units for Energy   7
   - The Unit of Toughness   9

2. **Handy Pressure Vessel Formulas**   11
   - Cylindrical Shells   11
   - Spherical Shell or Hemispherical Head   14
   - Elliptical Head   14
   - Torispherical Head   15
   - Conical Sections   17
   - Handy Formulas for Computing Head Weights   25
   - ASME F&D Head Weights   26
   - Hemispherical Head Weights   27
   - Partial Volumes and Pressure Vessel Calculations   28

3. **Dynamic Response of Pressure Vessels and Stacks**   41
   - Flow-Induced Vibration-Impeding Devices   56
   - Guy Cables, Remediation Devices, and Support of Flare Stacks   66

4. **Wind Loadings on Pressure Vessels and Stacks**   75
   - The Velocity Pressure Distribution, q_z 76
   - Wind Directionality Factor, K_d 76
   - Velocity Pressure Coefficient, K_z 77
   - Topographic Factor, K_{zt} 78
   - Basic Wind Speed, V 79
   - Importance Factor, I 79
   - Gust-Effect Factor, G 79
   - The Projected Area Normal to the Wind, A_f 82

5. Pressure Vessel Internal Assessment   91
    Forces on Internal Components   93
    Lined Plates and Internal Components   97
    Helpful Structural Formulations   98
    Internal Expansion Joints   109

6. Safety Considerations for Lifting and Rigging   111
    The Concept of a Ton Weighs Heavy When Lifting   111
    Maximum Capacity of Slings   112
    Bridles and Center of Gravity (CG)   116
    Lift Categories   116
    Preparing for the Lift   116
    American National Standard Institute (ANSI) Safety Codes   117

7. Lifting and Tailing Devices   119
    Impact Factor   120
    Tension at Net Section   122
    Hoop Tension—Splitting Failure Beyond Hole   124
    Double Plane Shear Failure   125
    Out-of-Plane Instability (Dishing) Failure   127
    Bearing Failure   129
    Pin Hole Doubler Plates   131
    Multiple Loads on Lifting and Tail Lugs   137
    Trunnions   150

8. Assessing Weld Attachments   163
    A Few Words About Reinforcing Pads and Lifting Lugs   171

9. Rigging Devices   181
    Blocks   181
    Selection of a Block   184
    Lifting and Erecting Pressure Vessels and Stacks   185
    Shackles   187

Index   189

A. Keith Escoe, P.E., has worked in the chemical process, oil refining, and pipeline industries for thirty-five years all over the world. His experience includes South America, North America, and the Middle East. He is retired from Saudi Aramco in the Kingdom of Saudi Arabia where he was an engineering specialist and returned to the United States as an international consultant. He is currently a Senior Principal Vessel Engineer with the Foster Wheeler USA Corporation in Houston, Texas. The author of many technical papers and books, Mr. Escoe has a B.S. in Mechanical engineering from the University of Texas at Arlington, an MBA from the University of Arkansas, and is a licensed professional engineer in Texas and a member of ASME. He participated with the Pressure Research Council for many years.

This book is meant to be a companion for those working in the field and facing various tasks involving pressure vessels and stacks. A typical example of these tasks is during a plant expansion project when equipment is modified for various reasons. One solution might be to increase the height of a particular process column to accommodate additional trays and/or packing to enhance process productivity. Another example is to retrofit existing pressure vessels or stacks to either enhance process capability or replace damaged equipment. Once these items are modified or constructed, it is up to the field personnel or hired contractors to devise mechanisms to lift them into place. These lifting devices (e.g., lifting and tailing lugs) become necessary to design and fabricate in order to install the desired pressure vessel or stack. Other cases may involve lifting devices which may have been removed and the equipment that must be moved for various reasons. There is no end to using lifting devices in an operations facility. This book does not get involved in the details of rigging, but does expand on the subject of various lift devices and useful rigging techniques.

Other issues face field personnel. Some may call these issues *problems*, but after reading this book it is hoped that these problems will be called *opportunities*. It has happened on many unfortunate occasions where process columns and stacks have been erected in the field and the structures vibrate. This involves a dynamic response induced from wind vortex shedding that results in disruption of the contained fluids and can be significant enough to be of concern for the safety of all present at the facility. Most people that observe such a dynamic response fear that the anchor bolts attaching the stack or column will fail due to static or fatigue loading on the anchor bolts. In this book, we discuss methods to screen and prevent such a reaction and practical methods to predict and correct the unstable motion should it occur.

Often defect mechanisms develop in portions of the process columns. To assess the stress, wind loads must be considered in the stress state at the location of the defect. Wind loads also become important when performing localized stress relieving and when a section of the process column is heated to stress relieving temperatures. When a section of a process column is to be stress relieved and heated up, wind striking the tower results in forces and moments across the heated section. This condition must be considered before repairs begin.

There is also a discussion about the application of guy cable supports for stacks in regards to dynamic response and wind loads. Of particular interest is a discussion about flare header stacks and how to design guy cables for these tall and slender structures.

Defect mechanisms also affect the internals of process columns. This book is a handy guide to the assessment of vessel internals and practical solutions. Tray support rings and catalyst bed support beams are some examples of vessel internals discussed.

Rather than present tables for conversion between U.S. Customary units and metric SI units, we begin with a section about unit systems. Before these topics are discussed, Chapter 1 discusses two unit systems: AES (American Engineering Units), or what the ASME calls the U.S. Customary Units, and the metric SI system of units. Because the United States is somewhat alone in the world using the AES, more foreign work and more foreign engineers are entering our country to practice engineering. Many foreign clients insist that the work be done using the metric SI system of units. All the calculations and the results are, by contract, to be in the metric SI system. Therefore, it is worthwhile spending some time on the two systems, because many of the examples are in one or the other system of units. This chapter will allow discussions and examples to be entirely in either system of units without converting from one to the other. Why not just pick up a basic text book on physics and read about metric SI? This is an option, but here the perspective is different. Our focus is on the units that one faces in field operations and how to clear up confusion dealing with fundamentals of mass and weight and also the various units of measurement encountered in the field. There is a detailed discussion about the two systems of units and how the various units are derived and used. This discussion is most important to read when it is necessary to work in a system of units one is not familiar with. It will also be a useful guide to refresh those who need to review the systems of units.

The examples in this book are from actual field applications. They come from various parts of the world and are written to enhance field operations. In many parts of the world, often in remote locations, these methods were applied to repair pressure vessels and stacks. These problems will still continue to happen, so there is a need to know how to address them. This book is to present assessments and techniques and methods for the repair of pressure vessels and stacks for field applications. Also the book is to be a repair manual for easy use for mechanical engineers, civil-structural engineers, plant operators, maintenance engineers, plant engineers and inspectors, materials specialists, consultants, and academicians.

There are also handy pressure vessel formulas—calculation of head formulas with partial loaded volumes and head weights—included, making this a handy field guide.

The contents of this book do not necessarily reflect the practices of my current employer, Foster Wheeler USA Corporation. I wish to thank J. Wesley Mueller, P.E., for his helpful comments pertaining to Chapters 6 and 7. I also wish to thank my wife, Emma, for her unrelenting patience throughout the project.

A. *Keith Escoe*
Houston, Texas
April 4, 2008

Chapter 1

Systems of Units

This chapter presents two systems of units so that you can follow the examples ahead. These two systems of units are the *metric SI* and what is termed by the American Society of Mechanical Engineers (ASME) as the *U.S. Customary system of units*, namely in the ASME Section II Part D [1]. This system is also termed the *American Engineering System (AES)* by the U.S. government. I mentioned the latter term in my book *Piping and Pipelines Assessment Guide* [2], in how to use the two systems of units. In this book, we will discuss briefly the other variants of the metric SI system, but it is the prevailing metric system of units. Likewise, we will concentrate on the U.S. Customary system versus the British Imperial system. Even though the latter two are similar, there are some differences.

This book is about engineering and discusses how to engineer with each system. It is not of interest to get into a historical discussion about how the system of units evolved, as there are many sources available if you have this interest. There are strong emotions associated with using each system, but this book is not interested in the polemics of using one system versus the other. The other reason for this discussion is that I have worked extensively in each system and have noticed the level of apprehension and intimidation among those using U.S. Customary units toward the metric SI system. This apprehension is totally unnecessary and is without warrant, as the metric SI is used in almost every country of the world except the United States, where it has made headway in medicine and the pure sciences. After reading this chapter, you will not need to convert from one system to the other in the discussions that follow; this text is for users of each system of units.

If you have used only the U.S. Customary system of units, the younger you are, the more likely you will be in the future to encounter the metric SI in practice. If you work outside the United States, then chances are certain that you will have to work in this system of units in one form or another. With more and more foreign projects and foreign engineers coming to the United States, the more likely the event of your using metric SI. Instead of resisting metric SI, consider it as a new friend, which it has been to me. In the metric SI, there are no fractions to worry about, like adding 3/32 to 11/64! The thought of not having to work with fractions is addictive in itself.

The metric SI is an absolute system of units, meaning that it does not depend on where the measure is made. The measurements can be made at any location. For example, the meter has the same (or absolute) length regardless of where the measurement is taken—here on earth or elsewhere. The unit of force is a derived unit. The metric SI system has been called the *meter, kilogram, and second* system, or *MKS*. These three units are primary units. In this system the Newton is the amount of force needed to give 1 Kg mass an acceleration of $1 \, m/sec^2$. Thus, Newton's second law is the crux of the system.

To derive *force* from *mass*, you have to use Newton's second law:

$$F = M*A, \text{Newtons} \qquad (1.1)$$

The unit of mass is kilogram (Kg) and acceleration is m/sec^2. To perform the conversion, you use

$$F = \left(\frac{g}{g_c}\right) M(Kg) \qquad (1.2)$$

In the metric SI system, you use

$$g = 9.807 \, \frac{m}{sec^2} \text{ and } g_c = 1.0 \, \frac{Kg - m}{N - sec^2}$$

Thus, the force required giving 1 Kg of mass an acceleration of $1 \, m/sec^2$ is

$$F = (1.0 \, Kg)\left(\frac{9.807 \, \dfrac{m}{sec^2}}{1\dfrac{Kg - m}{N - sec^2}}\right) = 9.807 \, N \qquad (1.3)$$

Yes, that's right: It takes 9.807 N (Newtons) to accelerate 1 Kg (kilogram) of mass $1 \, m/sec^2$—almost 10 times. This is a number to remember. See the note later in this section.

Regarding the U.S. Customary system, the same discussion is presented in my book *Piping and Pipelines Assessment Guide*, pp. 498–500 [2], as follows:

$$g = 32.174 \, \frac{ft}{sec^2} \text{ and } g_c = \frac{32.174 \, ft - lb_m}{sec^2 - lb_f}$$

From Newton's second law, we have the following:

$$\text{Force} = \frac{\text{Mass } (lb_m) * \text{Acceleration}\left(g = 32.174 \, \dfrac{ft}{sec^2}\right)}{\left(g_c = \dfrac{32.174 \, lb_m - ft}{lb_f - sec^2}\right)} = lb_f \qquad (1.4)$$

Hence,

$$\text{Mass} = \frac{\text{Force (lb}_\text{f}) * \left(g_c = \dfrac{32.174\ \text{lb}_\text{m} - \text{ft}}{\text{lb}_\text{f} - \text{sec}^2} \right)}{\text{Acceleration} \left(g = \dfrac{32.174\ \text{ft}}{\text{sec}^2} \right)} = \text{lb}_\text{m} \qquad (1.5)$$

As you can see, in the U.S. Customary system, mass is a derived unit, with the primary units being *force*, *pound*, and *second*. Some authors refer to it as the *FPS* system. This is a gravitational system, where force is a primary unit. Since most experiments involve a direct measurement of force, engineers prefer a gravitational system of units as opposed to an absolute system. Often the units g and gc are rounded to 32.2 ft/sec^2.

As you can see in equations 1.4 and 1.5, the terms lb_f and lb_m are used interchangeably. In the U.S. Customary system of units, lb$_\text{f}$ and lb$_\text{m}$ have the same magnitude (value). Pound (mass), lb$_\text{m}$, and pound (force), lb$_\text{f}$, have identical numerical values. Thus, 1 pound mass is equal to 1 pound force; hence, it is not uncommon to use the term *pound*, or *lb*, interchangeably. This usage has unfortunately caused confusion. Force is not mass, and this is hard to understand using the U.S. Customary system of units, where the same term is used for both mass and force. In locations without gravity, such as outer space, weight is meaningless. The official unit of mass in the U.S. Customary system of units is the slug. A pound is the force required to accelerate 1 slug of mass at 1 ft/sec^2. Since the acceleration of gravity in the U.S. Customary system is 32.2 ft/sec^2, it follows that the weight of one slug is 32.2 pounds, commonly referred to as 32.3 lb$_\text{m}$. The comparison of the slug and the pound makes it clear why the size of the pound is more practical for commerce. With the current scientific work, it is undesirable to have the weight of an object as a standard because the value of g does vary at different locations on Earth. It is much better to have a standard in terms of mass. The standard kilogram is the mass reference for scientific work. This book is for industrial practice by practicing engineers, inspectors, maintenance engineers, plant and pipeline personnel, rigging engineers, and others that work in industry. It is not intended for scientific work. The value of the gravitational constant does not vary enough to affect most engineering applications. The slug is rarely used outside of textbooks, which has contributed to the confusion between the pound mass and the pound force. When expressing mass in pounds, it is necessary to recognize that we are actually expressing "weight," which is a measure of the gravitational force on a body. When used in this manner, the weight is that of a mass when it is subjected to an acceleration of 1 g. In academia, where the study of dynamics involves forces, masses, and accelerations, it is important that mass be expressed in slugs, that is, m = W/g, where g is approximately equal to 32.174 ft/sec^2. These points are arguments for the use of the metric SI system of units. Particularly in the study of dynamics, the SI system is much easier. The slug is defined as

$$1 \, \text{Slug} = 32.174 \, \text{lb}_\text{m} = \frac{\text{lb}_\text{f} - \text{sec}^2}{\text{ft}} \tag{1.6}$$

Hence,

$$g_c = 1 = \frac{32.174 \, \text{lb}_\text{m} - \text{ft}}{\text{lb}_\text{f} - \text{sec}^2} \tag{1.7}$$

Like text books in academia, the slug is rarely used in industrial circles, but where it is used remember that $1 \, \text{slug} = 32.2 \, \text{lb}_\text{m}$. This will come up briefly in Chapter 3 where the ASME STS-1 uses the slug as mass. However, ASME is in the process of making the SI system the preferred system of units. In locations without gravity, such as outer space, weight is meaningless. If two bodies were to collide in outer space, the results would be due to their differences in mass and velocity. The body with the greater mass would win out.

   Note: Because lb_f and lb_m have the same unit and both are often referred to as *pounds*, it is a common mistake for users of the U.S. Customary system to forget to convert kilogram mass to Newton's force, or vice versa. When using the metric SI, don't forget the conversion factor of 9.807 derived earlier. Repeating again, kilograms are not Newtons. With the metric SI, this phenomenon does not exist, as 1 kilogram is 9.807 Newtons, so mass and weight cannot be confused.

GETTING FAMILIAR WITH METRIC SI UNITS

Civil-structural engineers prefer to work in units of force in designing foundations. With the U.S. Customary system, this is obvious: A pound is a pound. In the metric SI system, you must make a conversion. In the metric SI, KiloNewtons (KN) are used for foundations. Often I have heard the question "How do I convert kilograms to KiloNewtons?" The answer is simple. Suppose you have a pressure vessel that is a large reactor that is to go into a new refinery. This reactor weighs 1,000,000 kilograms. This is converted to force as follows:

$$F = 1,000,000 \, \text{Kg} * \frac{\left(9.807 \, \frac{\text{m}}{\text{sec}^2}\right)}{\frac{\text{Kg} - \text{m}}{\text{sec}^2 - \text{N}}} = 9,807,000 \, \text{N} \tag{1.8}$$

This means 1 million kilograms almost equals 10 million Newtons. So the civil-structural engineers would design for 9807 KiloNewtons (KN).

In Europe it is quite common to see lifting devices, such as small cranes in automobile shops, rated in KN. I saw a lifting crane in an automobile shop in Germany marked as 20 KN. This marking means that the crane could safely lift

$$20 \text{ KN} = 20,000 \text{ N}$$

Using Eq. 1.3, we have

$$\text{Mass (Kg)} = F * \left(\frac{g_c}{g} \right) = 20,000 \text{ N} * \left(\frac{1 \dfrac{\text{Kg} - \text{m}}{\text{N} - \text{sec}^2}}{9.807 \dfrac{\text{m}}{\text{sec}^2}} \right) = 2039.4 \text{ Kg}$$

So the crane is rated at roughly 2039 Kg. It would not be wise to ask how many pounds this is, as often many in the European Union are as emotional about the metric SI as some Americans are about the U.S. Customary system. In secret, you can calculate

$$2039.4 \text{ Kg} = 4496.1 \text{ lb}_m$$

Most people using the U.S. Customary system would then say the measurement is "4496.1 pounds."

If you are beginning to use the metric SI for the first time, it is quicker to learn the system by carrying all calculations solely in metric. This will enable you to become familiar with the system more quickly and obtain a "feel" for the answer.

OTHER IMPORTANT METRIC SI UNITS USED IN MECHANICS

The basic units—area, section modulus, and moment of inertia—are mm^2, mm^3, and mm^4, respectively.

Density

The density of steel is $0.283 \text{ lb}_m/\text{in}^3$. In the SI metric system, this measurement converts to approximately

$$\rho = 0.283 \frac{\text{lb}_m}{\text{in}^3} = 7840 \frac{\text{Kg}}{\text{m}^3}$$

Bending Moments and Torque

Because moment of force (bending moment) and torque are equal to a force times a distance (moment arm or lever arm), their SI unit is N*m. The Joule (J = N*m), which is a special name for the SI unit of energy and work, should not be used as a name for the unit of moment of force or of torque. Typically, the moment of torque is written as N m, with a space between the N and m or as N*m.

The Joule is equal to a 1 N*m, but is reserved for a unit of energy and can have more than one application, as discussed later. When we get into thermal stresses and heat transfer, it is confusing to use Joule as a bending moment of torque and as a thermal unit. We will spend more time later on the proper use of metric units.

UNITS FOR STRESS AND PRESSURE

The term for pressure and stress is *1 Newton per square meter*, which is named in honor of the famous mathematician, physicist, and philosopher Blaise Pascal. Since the area of a meter is rather large, 1,000 pascals is a kilopascal (KPa), and 1 million pascals is 1 megapascal (MPa). Simply written, we have

$$1.0 \text{ Megapascal} = (1,000,000)(1.0)\frac{N}{m^2}\left(\frac{1\,m}{1000\,mm}\right)^2 = 1.0\frac{N}{mm^2}$$

A megapascal, or MPa, is most commonly used for stress. It can also be used for pressure, but the numbers may remain small for small values of pressure. Typically, the kilopascal is used for pressure. The bar has been used for pressure often in the past, but the bar is not an SI unit. Although it may be accepted in the SI, it is discouraged. Now

$$1 \text{ kilopascal} = 0.01 \text{ bar} = 0.001 \text{ megapascal}$$

In U.S. Customary units, these metric units are

$$1 \text{ megapascal} = 145 \text{ psi} = 10 \text{ bars}$$
$$1 \text{ bar} = 14.5 \text{ psi}$$
$$1 \text{ kilopascal} = 0.145 \text{ psi}$$

A WARNING ABOUT COMBINING METRIC SI UNITS

When you are using the metric SI system of units, it is wise to remember that many units are named in terms of a magnitude of 10, e.g., *kilo* or *mega* as a prefix. When you are performing computations, it is advised to reduce these terms to their most basic set of units. For example, if you have a cylinder that is 609.6 mm (24″) ID that contains 1000 KPa of pressure that is 24 mm thick, the hoop stress is

$$\sigma = \frac{PD}{2t}$$

Entering the equation as

$$\sigma = \frac{(1575) \text{ KPa } (609.4) \text{ mm}}{2(24) \text{ mm}}$$

can lead to mistakes, since the stress term is in MPa and the pressure is in KPa. The best way to avoid mistakes is to write the equation as follows:

$$1575 \text{ KPa } (0.001) = 1.575 \text{ MPa} = 1.575 \frac{N}{\text{mm}^2}$$

If the pressure (written in KPa) is not converted correctly to megapascals (MPa), then big errors can occur. I have seen this problem occur with veteran users of the metric SI system. The error is even more likely with those not accustomed to using the metric SI system.

UNITS FOR ENERGY

The unit of heat is the Joule, mentioned previously. The Joule was named for James Prescott Joule, the famous English physicist. His development of Joule's Law, which related the amount of heat produced in a wire as proportional to the resistance of the wire and the square of the current, led to the thermal unit being named for him. The Joule is also used in Charpy impact tests, where an impact hammer is dropped from a specified height to impact a sample. The force is in Newtons, which is at a specified height in meters (or millimeters), and the energy that impacts the metal specimen is N*m or Joules. Refer to Figure 1.1, which shows a Charpy impact test machine.

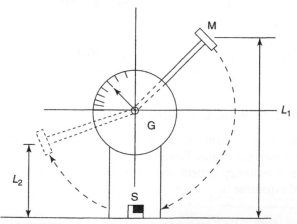

FIGURE 1.1   Schematic diagram showing impact hammer of M dropping from height L_1, impacting sample S, and rising to a height L_2. The energy absorbed by the sample, related to the difference of heights $L_1 - L_2$, is recorded on gauge G.

If the metal specimen does not break after impact, then it absorbed the energy of impact, which defines its toughness. We will discuss toughness later. The comparable impact energy used in the U.S. Customary system is the ft-lb$_f$. The unit ft-lb$_f$ can be used as an energy unit or as a bending moment of torque. The thermal unit in the U.S. Customary system is the British Thermal Unit, or BTU. For reference, 1 BTU approximately equals 1055 Joules, or 1.055 kJ (kilojoules).

The amount of thermal energy transferred per unit of time, power, is BTU/hr. In the SI metric, the comparable unit is Watt (W). Thus,

$$1\,\text{Watt} = 3.4128\,\frac{\text{BTU}}{\text{hr}} \text{ or } 1\,\frac{\text{BTU}}{\text{hr}} = 0.293\,\text{W}$$

The heat transfer convection coefficient in the U.S. system is BTU/(hr $-$ ft^2 $-$ °F). In the SI metric, the coefficient is Watts/(m^2 $-$ °C), or

$$1\,\frac{\text{Watt}}{\text{m}^2 - {}^\circ\text{C}} = 0.17612\,\frac{\text{BTU}}{\text{hr} - \text{ft}^2 - {}^\circ\text{F}}$$

or,

$$1\,\frac{\text{BTU}}{\text{hr} - \text{ft}^2 - {}^\circ\text{F}} = 5.68\,\frac{\text{Watt}}{\text{m}^2 - {}^\circ\text{C}}$$

Thermal Conductivity Units

The unit for thermal conductivity in the U.S. Customary system is BTU/(hr $-$ ft $-$ °F). Thus, in the SI metric,

$$1\,\frac{\text{Watt}}{\text{m} - {}^\circ\text{C}} = 0.57782\,\frac{\text{BTU}}{\text{hr} - \text{ft} - {}^\circ\text{F}}$$

or,

$$1\,\frac{\text{BTU}}{\text{hr} - \text{ft} - {}^\circ\text{F}} = 1.731\,\frac{\text{Watt}}{\text{m} - {}^\circ\text{C}}$$

Coefficient of Thermal Expansion

The U.S. Customary system unit for the coefficient of thermal expansion is microinch per inch per degree Fahrenheit. To convert to metric SI, you multiply the U.S. Customary system unit by 1.8. As an example, if the thermal coefficient of expansion is

$$\left(6.25 \times 10^{-6}\right)\frac{\text{in}}{\text{in} - {}^\circ\text{F}} = \left(1.125 \times 10^{-6}\right)\frac{\text{m}}{\text{m} - {}^\circ\text{C}}$$

A handy website for conversions is www.efunda.com.

THE UNIT OF TOUGHNESS

The unit of toughness is a very important parameter used in fracture mechanics. Toughness, K, is the property of a material to absorb energy. In the U.S. Customary system, this unit is expressed as $\text{ksi}\sqrt{\text{in}}$. When using the metric SI system, many people use $\text{MPa}\sqrt{\text{m}}$, mainly because it is closer in value to the U.S. Customary unit. The m denotes meters. The critical value of the mode stress intensity, K_I, at which fracture occurs is a function of the maximum uniform membrane stress. In the SI system, stress is usually denoted as MPa (N/mm^2). Since the stress unit MPa is $1.0 \ N/mm^2$, the unit for toughness becomes

$$\frac{lb_f}{in^2}\sqrt{in} = \frac{lb_f}{in^2}\left(\frac{4.448 \ N}{1 \ lb_f}\right)\left(\frac{in}{25.4 \ mm}\right)^2\left[in\left(\frac{25.4 \ mm}{1 \ in}\right)\right]^{0.5}$$

Thus,

$$1.0 \frac{lb_f}{in^2}\sqrt{in} = 0.0347 \frac{N\sqrt{mm}}{mm^2}$$

Since

$$1 \ MPa = 1\frac{N}{mm^2}, \ then$$

$$1.0 \ MPa\sqrt{mm} = 28.78 \ psi \sqrt{in}$$

or,

$$1.0 \ MPa\sqrt{mm} = 0.02878 \ ksi \sqrt{in}$$

If you use meters instead of millimeters, the preceding becomes

$$1.0 \ MPa\sqrt{\frac{mm\,(m)}{(1000 \ mm)}} = 0.03162 \ MPa\sqrt{m}$$

Thus,

$$1 \ MPa\sqrt{m} = 0.91 \ ksi\sqrt{in}$$

Since these units are fairly close, many people prefer to use $MPa\sqrt{m}$. The same argument can be used for using centimeters versus millimeters—the centimeter is closer to 1 inch than the millimeter; however, all dimensions are given in millimeters in countries that use the metric SI. Thus, you can use the unit $MPa\sqrt{mm}$ or $MPa\sqrt{m}$, although the former is more consistent with the dimensions, which are in millimeters. Either unit is acceptable, as long as you keep the units consistent.

REFERENCES

1. ASME Section II Part D, Properties Materials, American Society of Mechanical Engineers, New York, NY, 2007.
2. Escoe, A. Keith, *Piping and Pipelines Assessment Guide*, Gulf Professional Publishing (Elsevier), March 2006.

Chapter 2

Handy Pressure Vessel Formulas

This chapter contains handy formulas for pressure vessels. Some of the formulas are from ASME, Section VIII, Division 1 [1], and others are associated formulations to calculate weights and partial fluid volumes.

In field applications, it is assumed that the equipment has already been fabricated and been shop-tested for the maximum allowable pressure (MAP). The MAP is defined as the maximum allowable pressure of the vessel in the new and cold condition. It is more often determined in the shop before delivery. After the vessel is delivered, any test performed after operation begins is the maximum allowable working pressure (MAWP). The MAWP can also be used for new construction. The maximum allowable working pressure is defined as the maximum gauge pressure permissible at the top of the completed vessel in its operating condition for a designated temperature. Thus, in the field, you are likely to hear the term MAWP much more than MAP.

The minimum required wall thickness for a component can be taken as the thickness in the new condition minus the original specified corrosion allowance. The minimum required wall thickness for pressure vessel components can be computed if the component geometry, design pressure (including liquid head) and temperature, specifications for the material of construction, allowable stress, and thicknesses required for supplemental loads are known. The values for thickness calculations must include future corrosion allowance—the amount of corrosion expected after several field inspections are performed. Refer to the API 579, "Fitness-for-Service" [2], for additional discussion.

CYLINDRICAL SHELLS

Three formulas that are always helpful in mechanics problems are the properties of area, section modulus, and moment of inertia for the cross-section of a circular cylinder (see Figure 2.1). These formulas are as follows, with the approximate formulations on the left and the exact expressions on the right side:

$$I = \pi R^3 t; \; Exact \; I = \frac{\pi}{64}(D_o^4 - D_i^4) \tag{2.1}$$

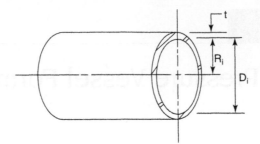

FIGURE 2.1   Right circular cylinder.

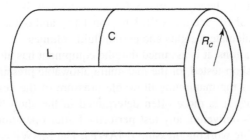

FIGURE 2.2   Right circular cylinder showing circumferential and longitudinal axes.

$$Z = \pi R^2 t; Exact = Z = \frac{\pi}{32}\left(\frac{D_o^4 - D_i^4}{D_o}\right) = \left(\frac{2}{D_o}\right)I \tag{2.2}$$

$$A = \pi R t; Exact = A = \frac{\pi}{4}\left(D_o^2 - D_i^2\right) \tag{2.3}$$

where

A = cross-sectional area of cylindrical shell, mm^2 (in^2)
D_o = outside diameter of cylindrical shell, mm (in)
D_i = inside diameter of cylindrical shell, mm (in)
I = moment of inertial of cylindrical shell cross-section, mm^4 (in^4)
R = mean radius of cylindrical shell in approximate formulation, mm (in)
t = thickness of cylindrical shell in approximate formulation, mm (in)
Z = section modulus of cylindrical shell cross-section, mm^3 (in^3)

In the era of high-speed computers, there is no reason for the exact expressions not to be used.

Circumferential Stress in a Cylindrical Shell (Longitudinal Joints)

The equations for a right circular cylinder for the circumferential stress acting along the longitudinal joints follow (see Figure 2.2).

$$t_r^C = \frac{PR_C}{S_a E - 0.6P} \qquad (2.4)$$

$$MAWP^c = \frac{S_a E t_c}{R_C - 0.6t_C} \qquad (2.5)$$

$$\sigma_m^C = \frac{P}{E}\left(\frac{R_C}{t_C} + 0.6\right) \qquad (2.6)$$

where

E = weld joint efficiency from original construction code; if unknown, use 0.7
$MAWP$ = maximum allowable working pressure, MPa (psi)
P = internal design pressure, MPa (psi)
$R_C = R + LOSS + FCA$
R = Inside radius, mm (in)
$LOSS$ = wall loss in the shell prior to the assessment equal to the nominal (or furnished thickness if available) minus the measured minimum thickness at the time of the inspection, mm (in)
FCA = Future corrosion allowance—the amount of wall loss expected over the specified time of the assessment predicting the remaining life based on inspection data or estimates, mm (in)
S_a = allowable tensile stress of the shell material evaluated at the design temperature per the applicable construction code, MPa (psi)
$t_C = t - LOSS - FCA$, mm (in)
t = nominal or furnished thickness of the shell, or cylinder thickness at a conical transition for a junction reinforcement calculation, mm (in)
t_r = required minimum wall thickness
σ_m = nominal membrane stress

Longitudinal Stress (Circumferential Joints)

The equations for a right circular cylinder for the longitudinal stress acting on the circumferential joints follow.

$$t_r^L = \frac{PR_C}{2S_a E + 0.4P} + t_{sl} \qquad (2.7)$$

$$MAWP^L = \frac{2S_a E\left(t - t_{sl}\right)}{R_C - 0.4\left(t_C - t_{sl}\right)} \qquad (2.8)$$

$$\sigma_m^L = \frac{P}{2E}\left(\frac{R_C}{t_C - t_{sl}} - 0.4\right) \qquad (2.9)$$

where t_{sl} = thickness required by supplemental loads, e.g., wind or seismic loads, mm (in).

Final or Resulting Values

$$t_r = \max\left(t_r^C, t_r^L\right) \tag{2.10}$$

$$MAWP = \min\left(MAMP^C, MAWP^L\right) \tag{2.11}$$

$$\sigma_{\max} = \max\left(\sigma_m^C, \sigma_m^L\right) \tag{2.12}$$

SPHERICAL SHELL OR HEMISPHERICAL HEAD

The equations for a spherical shell or hemispherical head follow (see Figure 2.3).

The minimum thickness, MAWP, and equations for the membrane stress are given in the ASME, Section VIII, Division 1, Boiler and Pressure Vessel Code, paragraph UG-27 [1], as follows:

$$t_{\min} = \frac{PR_C}{2S_a E - 0.2P} \tag{2.13}$$

$$MAWP = \frac{2S_a E t_C}{R_C + 0.2t_C} \tag{2.14}$$

$$\sigma_m = \frac{P}{2E}\left(\frac{R_C}{t_C} + 0.2\right) \tag{2.15}$$

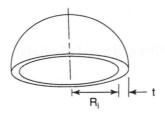

R_i t

FIGURE 2.3   Hemispherical head or sphere.

ELLIPTICAL HEAD

The equations for an elliptical head follow (see Figure 2.4).

The minimum thickness, MAWP, and membrane stress equations are as follows per the ASME, Section VIII, Division 1, Appendix 1 [1] code:

$$t_{\min} = \frac{PD_C K}{2S_a E - 0.2P} \tag{2.16}$$

$$MAWP = \frac{2S_a E t_C}{KD_C + 0.2t_C} \tag{2.17}$$

$$\sigma_m = \frac{P}{2E}\left[\frac{D_C K}{t_C} + 0.2\right] \tag{2.18}$$

where

$$D_C = 2R_c \tag{2.19}$$

$$K = \frac{1}{6}\left(2.0 + R_{\text{ell}}^2\right) \tag{2.20}$$

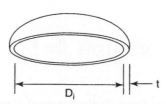

FIGURE 2.4 Ellipsoidal head.

R_{ell} = Ratio of the major-to-minor axis of an elliptical head (most common is $R_{\text{ell}} = 2$ for a 2:1 ellipsoidal head)

Note: To compute the minimum thickness, MAWP, and membrane stress for the spherical portion of an ellipsoidal head, defined as a section within 0.8D centered on the head centerline, use K_c instead of K in the preceding equations. K_c is defined as follows:

$$K_c = 0.25346 + 0.13995R_{\text{ell}} + 0.12238R_{\text{ell}}^2 - 0.015297R_{\text{ell}}^3 \tag{2.21}$$

TORISPHERICAL HEAD

The equations for a torispherical head follow (see Figure 2.5).

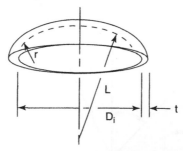

FIGURE 2.5 Torispherical head (**Note:** $Cr = L$ below).

$C_{rc} = C_r + \text{Loss} + \text{FCA (mm, in)}$
$R_c = R + \text{Loss} + \text{FCA (mm, in)}$

The minimum thickness, MAWP, and membrane stress equations are as follows:

$$t_r = \frac{PC_{rc}M}{2SE - 0.2P} \tag{2.22}$$

$$MAWP = \frac{2S_aEt_c}{C_{rc}M + 0.2t_c} \tag{2.23}$$

$$\sigma_m = \frac{P}{2E}\left(\frac{C_{rc}M}{t_c} + 0.2\right) \tag{2.24}$$

where

$$M = \frac{1}{4}\left(3.0 + \sqrt{\frac{C_{rc}}{r}}\right) \tag{2.25}$$

Geometrical Equations for a Torispherical Head

Referring to Figure 2.6, we have the following:

$$\alpha = \arcsin\left(\frac{R_c}{C_{rc}}\right) \tag{2.26}$$

In most cases, $49° \leq \varphi \leq 65°$, depending on the thickness and diameter of the head. For many cases $\varphi \approx 55°$.

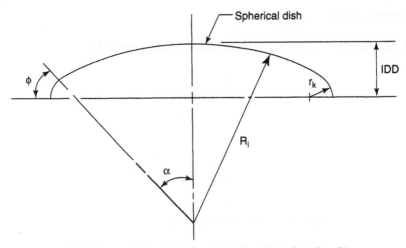

FIGURE 2.6   Torispherical head geometry (where $R_i = C_r = L$).

The equation for the knuckle angle is as follows:

$$\varphi = \arccos\left[\frac{IDD - C_r(1 - \cos(\alpha))}{r_k}\right]\left(\frac{180}{\pi}\right) \qquad (2.27)$$

CONICAL SECTIONS

The equations for conical sections follow, referring to Figures 2.7, 2.8, and 2.9.

Circumferential Stress (Longitudinal Joints)

$$t_r^C = \frac{PD_c}{2\cos\alpha\left(S_a E - 0.6P\right)} \qquad (2.28)$$

where $D_c = D + 2(LOSS) + FCA$

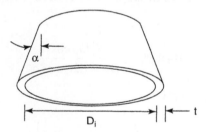

FIGURE 2.7   Conical section.

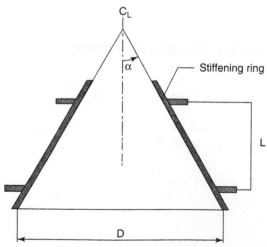

FIGURE 2.8   Conical transition section (Courtesy of the American Petroleum Institute).

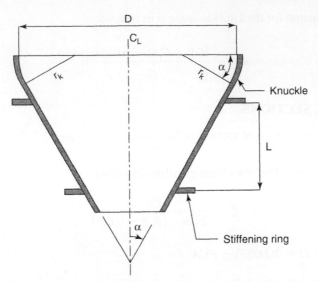

FIGURE 2.9 Toriconical head geometry (Courtesy of the American Petroleum Institute).

$$MAWP^C = \frac{2S_a Et_c \cos \alpha}{D_c + 1.2t_c \cos \alpha} \tag{2.29}$$

where $t_c = t - LOSS - FCA$

$$\sigma_m^c = \frac{P}{2E}\left(\frac{D_c}{t_c \cos \alpha} + 1.2\right) \tag{2.30}$$

where $R_c = R + LOSS + FCA$

Longitudinal Stress (Circumferential Joints)

$$t_r^L = \frac{PD_c}{2 \cos \alpha \left(2S_a E + 0.4P\right)} + t_{sl} \tag{2.31}$$

where t_{sl} = thickness required for any supplemental load based on the longitudinal stress, such as weight, wind, or seismic loads.

$$MAWP^L = \frac{4S_a E \left(t_c - t_{sl}\right) \cos \alpha}{D_c - 0.8 \left(t_c - t_{sl}\right) \cos \alpha} \tag{2.32}$$

$$\sigma_m^L = \frac{P}{2E}\left(\frac{R_c}{2 \left(t_c - t_{sl}\right) \cos \alpha} - 0.4\right) \tag{2.33}$$

Final Values

$$t_r = \max\left(t_r^C, t_r^L\right) \tag{2.34}$$

$$MAWP = \min\left(MAWP^C, MAWP^L\right) \tag{2.35}$$

$$\sigma_{\max} = \max\left(\sigma_m^C, \sigma_m^L\right) \tag{2.36}$$

Knuckle Section

$$t_r^k = \frac{PL_{kc}M}{2S_aE - 0.2P} \tag{2.37}$$

where

$$L_{kc} = \frac{R_c - r_{kc}\left(1 - \cos\alpha\right)}{\cos\alpha} \tag{2.38}$$

$$M = \frac{1}{4}\left(3.0 + \sqrt{\frac{L_{kc}}{r_{kc}}}\right) \tag{2.39}$$

Final Values

$$t_r = \max\left(t_r^c, t_r^k\right) \tag{2.40}$$

$$MAWP = \min\left(MAWP^C, MAWP^L\right) \tag{2.41}$$

$$\sigma_{\max} = \max\left(\sigma_m^c, \sigma_m^k\right) \tag{2.42}$$

Conical Transitions

The minimum thickness, MAWP, and membrane stress equations are computed on a component basis. You can use the preceding equations to compute the minimum required thickness, MAWP, and membrane stress of the cone section. These values are designated as t_r^c, $MAWP^C$, and σ_m^c, respectively. These parameters are shown in Figure 2.10.

Conical Transition Knuckle Section

You can use the following equations to compute the required thickness, MAWP, and membrane stress for the knuckle region, if used:

$$t_r^k = \frac{PL_{kc}M}{2S_aE - 0.2P} \tag{2.43}$$

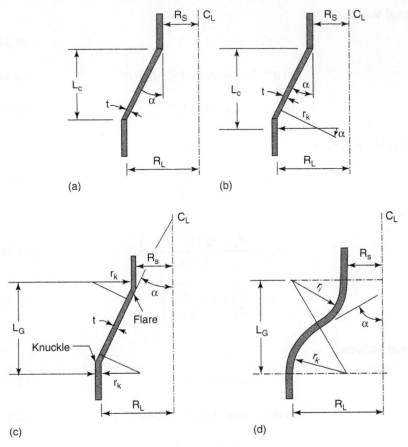

Note: $r_k \Rightarrow \max[0.12(R_L + t), 3t_c]$: R_S has no dimensional requirements.

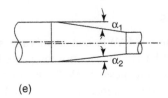

(e)

$\alpha_1 > \alpha_2$: Therefore use α_1 in design equations.

FIGURE 2.10 Conical transition geometry (courtesy of the American Petroleum Institute).

$$MAWP^k = \frac{2S_a Et_{kc}}{L_{kc}M + 0.2t_{kc}} \tag{2.44}$$

$$\sigma_m^k = \frac{P}{2E}\left(\frac{L_{kc}M}{t_{kc}} + 0.2\right) \tag{2.45}$$

where

$$L_{kc} = \frac{R_{LC} - r_{kc}\,(1 - \cos\alpha)}{\cos\alpha} \tag{2.46}$$

$$M = \frac{1}{4}\left(3.0 + \sqrt{\frac{L_{kc}}{r_{kc}}}\right) \tag{2.47}$$

Conical Transition Flare Section

$$t_r^f = \frac{PL_{fc}M}{2S_a E - 0.2P} \tag{2.48}$$

$$MAWP^f = \frac{2S_a Et_{fc}}{L_{fc}M + 0.2t_{fc}} \tag{2.49}$$

$$\sigma_m^f = \frac{P}{2E}\left(\frac{L_{fc}M}{t_{fc}} + 0.2\right) \tag{2.50}$$

where

$$r_{fc} = r_k + LOSS + FCA \tag{2.51}$$

$$L_{fc} = \frac{R_{Sc} + r_{fc}\,(1 - \cos\alpha)}{\cos\alpha} \tag{2.52}$$

$$M = \frac{1}{4}\left(3.0 + \sqrt{\frac{L_{fc}}{r_{fc}}}\right) \tag{2.53}$$

Equations Based on a Pressure-Area Force Balance Procedure

$$t_r^f = \left(\frac{1}{r_{fc}}\right)\left(\frac{P\,[K_1 + K_2 + K_3]}{1.5S_a E} - K_4 - K_5\right) \tag{2.54}$$

$$MAWP^f = 1.5S_a E \left(\frac{t_{fc}\alpha_r r_{fc} + K_4 + K_5}{K_1 + K_2 + K_3} \right) \qquad (2.55)$$

$$\sigma_m^f = \frac{P(K_1 + K_2 + K_3)}{1.5\, E\, (t_{fc}\alpha_r r_{fc} + K_4 + K_5)} \qquad (2.56)$$

where

$$K_1 = 0.125\, (2r_{fc} + D_1)^2 \tan \alpha - \frac{\alpha_r r_{fc}^2}{2} \qquad (2.57)$$

$$K_2 = 0.28 D_1 \sqrt{D_1 t_c^s} \qquad (2.58)$$

$$K_3 = 0.78 K_6 \sqrt{K_6 t_c^c} \qquad (2.59)$$

$$K_4 = 0.78 t_c^c \sqrt{K_6 t_c^c} \qquad (2.60)$$

$$K_5 = 0.55 t_c^s \sqrt{D_1 t_c^s} \qquad (2.61)$$

$$K_6 = \frac{D_1 + 2r_{fc}(1 - \cos \alpha)}{2 \cos \alpha} \qquad (2.62)$$

$$\alpha_r = \alpha \left(\frac{\pi}{180} \right) \qquad (2.63)$$

$$D_1 = 2R_s \qquad (2.64)$$

where
t^c = nominal or furnished small-end cone thickness in a conical transition
$t_c^c = t^c - LOSS - FCA$
t^s = nominal or furnished small-end cylinder thickness in a conical transition
$t_c^s = t^s - LOSS - FCA$

Final Values

Case 1: The conical transition contains only a cone; see Figure 2.11(a).

$$t_r = t_r^c \qquad (2.65)$$

$$MAWP = MAWP^c \qquad (2.66)$$

$$\sigma_{max} = \sigma_m^c \qquad (2.67)$$

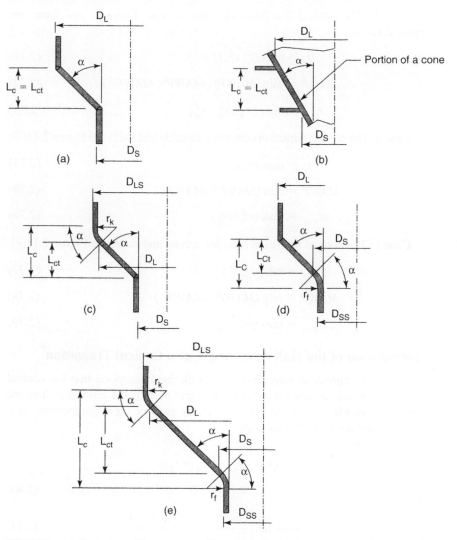

FIGURE 2.11 Conical transition geometry—Unsupported length for conical transitions (courtesy of the American Petroleum Institute).

Case 2: The conical transition contains a cone and knuckle; see Figure 2.11(b).

$$t_r = \max (t^c_{\min}, t^k_{\min}) \qquad (2.68)$$

$$MAWP = \min (MAWP^c, MAWP^k) \qquad (2.69)$$

$$\sigma_{\max} = \max (\sigma^c_m, \sigma^k_m) \qquad (2.70)$$

Case 3: The conical transition contains a cone, knuckle, and flare; see Figure 2.11(c).

$$t_r = \max\,(t_r^c, t_r^k, t_r^f) \tag{2.71}$$

$$MAWP = \min\,(MAWP^c, MAWP^k, MAWP^f) \tag{2.72}$$

$$\sigma_{max} = \max\,(\sigma_m^c, \sigma_m^k, \sigma_m^f) \tag{2.73}$$

Case 4: The conical transition contains a knuckle and flare; see Figure 2.11(d).

$$t_r = \max\,(t_r^k, t_r^f) \tag{2.74}$$

$$MAWP = \min\,(MAWP^k, MAWP^f) \tag{2.75}$$

$$\sigma_{max} = \max\,(\sigma_m^k, \sigma_m^f) \tag{2.76}$$

Case 5: The conical transition contains a cone and flare; see Figure 2.11(e).

$$t_r = \max\,(t_r^c, t_r^f) \tag{2.77}$$

$$MAWP = \min\,(MAWP^c, MAWP^f) \tag{2.78}$$

$$\sigma_{max} = \max\,(\sigma_m^c, \sigma_m^f) \tag{2.79}$$

Computation of the Half-Apex Angle of a Conical Transition

The following equations were developed with the assumption that the conical transition contains a cone section, knuckle, and flare. If the transition does not contain a knuckle or flare, you should set the radii of these components to 0 when computing the half-apex angle.

If

$$(R_L - r_k) > (R_S + r_f):$$

$$\alpha = \beta + \phi \tag{2.80}$$

$$\beta = \arctan\left[\frac{(R_L - r_k) - (R_S + r_f)}{L_c}\right] \tag{2.81}$$

If

$$(R_L - r_k) < (R_S + r_f):$$

$$\alpha = \beta - \phi \tag{2.82}$$

$$\beta = \arctan\left[\frac{(R_S + r_f) - (R_L - r_k)}{L_c}\right] \tag{2.83}$$

with

$$\phi = \arcsin\left[\frac{(r_f + r_k)\cos\beta}{L_c}\right] \tag{2.84}$$

HANDY FORMULAS FOR COMPUTING HEAD WEIGHTS

Ellipsoidal, flanged and dished (F&D), and hemispherical heads are made simply with a blank that has a diameter larger than the finished part and is formed by spinning or using an alternate forming process. During forming, the wall thickness is carefully controlled. The resulting product is formed through the use of forming rollers with specific profiles that are set at precise distances from each other and the mandrel. In one process, a complex shape is formed by a flat blank that is "sheared" by one or more rollers over a rotating mandrel. There is no material lost in the process. Such a process is shown in Figure 2.12.

2:1 ellipsoidal head weights

The weight of a head can be quickly and accurately found by computing the volume of a circular blank. For a 2:1 ellipsoidal head, the blank equation is

$$BD = 1.22\,(ID) + 2\,(S.F.) + T \tag{2.85}$$

where

BD = Blank diameter, in (mm)
ID = Inside diameter of head, in (mm)
S.F. = Straight flange of head, in (mm)
T = Head (or blank) thickness, in (mm)

FIGURE 2.12   Forming a formed head in the mill.

EXAMPLE 2.1

Find the weight of a 2:1 ellipsoidal head that has an inside diameter of 78 inches, is ⅜ inch thick, and has a straight flange of 2 inches.

    Solution:

$$BD = 1.22(78)\ \text{in} + 2(2.0)\ \text{in} + \frac{3}{8}\ \text{in} = 99.535\ \text{in}$$

Now you can compute the weight by multiplying the volume of the blank by the density of steel, as follows:

$$\left(\rho = 0.283\,\frac{\text{lb}_m}{\text{in}^3} = 489.02\,\frac{\text{lb}_m}{\text{ft}^3} = 7834\,\frac{\text{Kg}}{\text{m}^3} = 7.833\,\frac{\text{g}}{\text{cm}^3}\right),$$

$$\text{Wgt} = (0.283)\,\frac{\text{lb}_m}{\text{in}^3}\left(\frac{\pi}{4}\right)(99.535)^2\ \text{in}^2\,(0.375)\,\text{in} = 825.8\,\text{lb}_m$$

Thus, the head weighs $825.8\,\text{lb}_m$, which agrees with the steel mill's catalog.

ASME F&D HEAD WEIGHTS

The blank equation for an ASME F&D head is

$$BD = OD + 2\,(ICR) + T \tag{2.86}$$

where

ICR = Inside crown radius = r_k in Figure 2.6. Many people use the term *IKR* for inside knuckle radius.
OD = Outside diameter of head, in (mm)
T = Thickness of blank, in (mm)
OD = Outside diameter of blank, in (mm)

EXAMPLE 2.2

Compute the weight of an ASME F&D head that has an ID of 78 inches, a thickness of ⅜ inch, and a knuckle radius of 4¾ inches.

    Solution:

$$OD = 78\,\text{in} + 2\left(\frac{3}{8}\right)\text{in} = 78.75\,\text{in}$$

$$BD = 78.75\,\text{in} + 2(4.75)\,\text{in} + \frac{3}{4}\,\text{in} = 88.625\,\text{in}$$

$$Wgt = (0.283)\frac{\text{lb}_\text{m}}{\text{in}^3}\left(\frac{\pi}{4}\right)(88.625)^2\,\text{in}^2\left(\frac{3}{8}\right)\text{in} = 654.67\,\text{lb}_\text{m}$$

The manufacturer's catalog lists the head weight as being $654\,\text{lb}_\text{m}$.
  To compute the values of α and ρ, you may use Eq. 2.26 and Eq. 2.27, respectively.

HEMISPHERICAL HEAD WEIGHTS

You might think that hemispherical head weights can be computed easily from the following formulation:

$$Wgt = \frac{2\pi}{3}(R_\text{o}^3 - R_\text{i}^3)\,\text{in}^3(0.283)\frac{\text{lb}_\text{m}}{\text{in}^3} \qquad (2.87)$$

The truth is that the blanks used normally produce a head approximately 11% higher in weight than that computed with Eq. 2.87. If the heads are forged, which they are for very thick wall heads, then Eq. 2.87 is more accurate. Blanks can be used up to about 6 inches. Forging is used for thicker wall heads. Six inches is thick, so forgings can start at any thickness, particularly over 2 inches of wall thickness. To correct for this error for heads formed from blanks, you can use the following formulas:

$$BD = 1.506\,(\text{ID}) + T \qquad for\ T \leq 2\ \text{in} \qquad (2.88)$$

$$BD = 1.506\,(\text{ID}) + T\left(\frac{T}{2\ \text{in}}\right)^{0.03} \qquad for\ T > 2\ \text{in} \qquad (2.89)$$

where

BD = Blank diameter, in
ID = Inside diameter of head, in
T = Thickness of head wall, in

Note: These equations are empirical and developed using U.S. Customary units. If you are using the metric SI system, it is recommended that you use the equations with U.S. Customary units and then convert them to the metric SI system.

EXAMPLE 2.3

Consider a 78-inch ID hemispherical head with a nominal 1-inch wall. Compute the weight of the head.

From Eq. 2.88, you can compute the diameter of the blank as follows:

$$BD = 1.506(78) + 1 = 118.5 \text{ in}$$

The computed weight is as follows:

$$Wgt = \left(\frac{\pi}{4}\right)(78)^2 \text{ in}^2 \, (1) \text{ in} \, (0.283)\frac{lb_m}{in^3} = 3119.46 \, lb_m$$

Using the actual blank that is used in the mill gives the following:

$$BD = 118.5 \text{ in}$$

The error between the actual mill weight and that computed from Eq. 2.88 is 0.054%. The error between the actual mill weight and that computed from Eq. 2.87 is 11.11 %.

EXAMPLE 2.4

Compute the weight of a hemispherical head with an ID of 132 inches and a minimum head thickness of 6 inches.

Applying Eq. 2.89, you can determine the blank diameter as follows:

$$BD = 1.506(132) + 6\left(\frac{6}{2}\right)^{0.03} = 204.993 \text{ in}$$

$$Wgt = \left(\frac{\pi}{4}\right)(204.993)^2 \text{ in}^2 \, (6) \text{ in} \, (0.283)\frac{lb_m}{in^3} = 56041.124 \, lb_m$$

The actual blank diameter used at the mill is 207.5 inches in diameter. Therefore, you can calculate the weight as follows:

$$Wgt_{act} = \left(\frac{\pi}{4}\right)(207.5)^2 \text{ in}^2 (6) \text{in}\left(\frac{6 \text{ in}}{2 \text{ in}}\right)^{0.03} (0.283)\frac{lb_m}{in^3} = 57420.211 \, lb_m$$

The error between the results of Eq. 2.8 and the actual weight is 2.40%. The error between the actual weight and the result of Eq. 2.87 is 11.48%.

PARTIAL VOLUMES AND PRESSURE VESSEL CALCULATIONS

Listed in the following are formulations for the volumes of liquids occupying partial volumes.

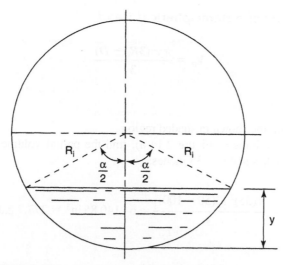

FIGURE 2.13 Sketch for calculating partial volume of a cylinder.

Partial Volume of Cylinder in Horizontal Position

The partial volume of a liquid in a horizontal circular cylinder, referring to Figure 2.13, is

$$V_P = \frac{R_i^2 L}{2}\left(\frac{\pi\alpha^o}{180} - \sin\alpha\right) = \text{partial volume shown in Figure 2.13} \quad (2.90)$$

where

L = length of cylinder
R_i = inside radius of cylinder

EXAMPLE 2.5

For a cylinder with a 144-inch ID, find the partial volume of a liquid head of 60 inches, if L = 100 ft.

$$\frac{\alpha}{2} = 80.41°$$

$$V_P = \frac{(72)^2(1200)}{2}\left[\frac{\pi(160.81)}{180} - \sin(160.81°)\right]$$

$$V_P = 7{,}707{,}650.2\,\text{in}^3 = 33{,}366.5\,\text{gal}$$

Partial Volume of a Hemispherical Head

$$V_P = \frac{\pi y^2 (3R_i - D)}{3} \qquad (2.91)$$

where

V_P = partial volume shown in shaded region

For vertical volume in Figure 2.14(a), find the partial volume jfor a head with $R_i = 50$ inches and $y = 35$ inches:

$$V_P = \frac{\pi (35)^2 \left[3(50 - 100) \right]}{3} = 64{,}140.85 \text{ in}^3 = 277.7 \text{ gal}$$

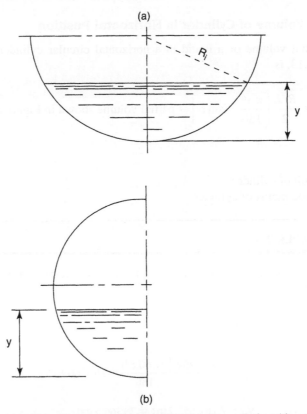

FIGURE 2.14 **(a)** Partial volume of a hemispherical head in the vertical position. **(b)** Partial volume of a hemispherical head in the horizontal position.

EXAMPLE 2.6

For horizontal volume in Figure 2.14(b), find the partial volume for a head
with $R_i = 50$ inches and $y = 35$ inches:

$$V_p = \frac{277.7}{2} = 138.85\,\text{gal}$$

Partial Volumes of Spherically Dished Heads

The equations for partial volumes of a liquid occupying spherically dished
heads follow.

Spherically Dished Head in Horizontal Position

The partial volume of a horizontal head shown in Figure 2.15 is

$$V = \alpha \left| \frac{\sqrt{\left(\rho^2 - y_i^2\right)^3} - \sqrt{\left(\rho^2 - R_i^2\right)^3}}{3} - \frac{L\left(R_i^2 - y_i^2\right)}{2} \right| \qquad (2.92)$$

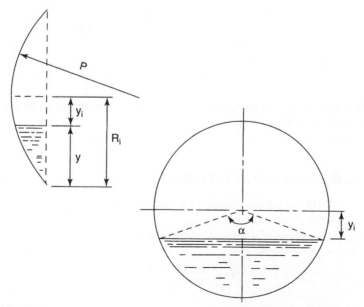

FIGURE 2.15   Partial volume of a spherically dished head in the horizontal position.

EXAMPLE 2.7

A spherically dished head with a 114-inch OD is spun from a 1-inch plate. Determine the partial volume of liquid that is at the bottom portion of the head. The head is shown in Figure 2.16, with a liquid level 10.0 inches below the centerline. From the vessel head manufacturer's catalog, you can determine the following:

$$IDD = 16.786 \text{ in}, \, \rho = 108 \text{ in}$$

$$R_i = \frac{114 - 2(1.0)}{2} = 56.0 \text{ in}$$

$$\alpha = 159.43° = 2.78 \text{ radians}$$
$$L = 108 - 16.786 = 91.21 \text{ in}$$

$$V = 2.78 \left| \frac{\sqrt{(108^2 - 10^2)^3} - \sqrt{(108^2 - 56^2)^3}}{3} - \frac{(91.21)(56^2 - 10^2)}{2} \right|$$

$$V = 37,677.63 \text{ in}^3 = 163.1 \text{ gal}$$

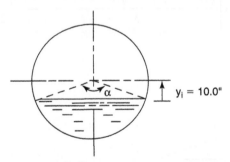

FIGURE 2.16 Spherically dished head in the horizontal position showing the liquid level at 10.0 inches below the centerline.

Spherically Dished Head in Vertical Position

The equation for the partial volume of a liquid occupying a spherically dished head in the vertical position is

The partial volume of a vertical head in Figure 2.17 is

$$V = \frac{\pi y(3x^2 + y^2)}{6} \tag{2.93}$$

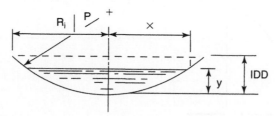

FIGURE 2.17 Volume of a spherically dished head in the vertical position.

or

$$V = \frac{\pi y^2 (3\rho - y)}{3} \tag{2.94}$$

EXAMPLE 2.8

For the same head in the example in Figure 2.16, determine the partial volume of a head of liquid of 9 inches.

$$x = 55.456 \text{ in.}$$
$$V = \frac{\pi(9)\left[3(55.456)^2 + 9^2\right]}{6} = 14,874 \text{ in}^3 = 64.4 \text{ gal}$$

Partial Volumes of Elliptical Heads

The partial volumes of elliptical heads are described in the horizontal and vertical positions as follows:

Elliptical Head in Horizontal Position

The exact partial volume of a horizontal elliptical head, as illustrated in Figure 2.18, is:

$$V = \frac{(\text{IDD})\alpha}{3R_i} \sqrt{(R_i^2 - y_i^2)^3} \tag{2.95}$$

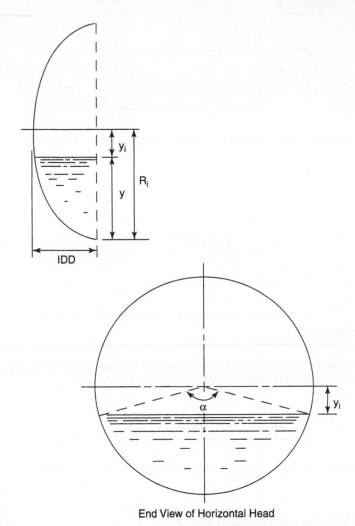

End View of Horizontal Head

FIGURE 2.18　Partial volume of an ellipsoidal head in the horizontal position showing the elevation and front views.

EXAMPLE 2.9

Find the partial volume of a 2:1 elliptical head (R_i/IDD = 2) for which the OD is 108 inches. The level of the liquid is 35 inches and the head is spun from a 1-inch plate.

$$IDD = \frac{108 - 2(1.0)}{4} = 26.50 \text{ in}$$

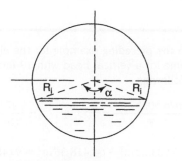

FIGURE 2.19 Ellipsoidal head in horizontal position example.

From Eq. 2.95 and Figure 2.19, you get the following:

$$V = \frac{(IDD)\alpha}{3R_i} \sqrt{(R_i^2 - y_i^2)^3}$$

$$\alpha = 138.80° = 2.42 \text{ radians}$$

$$V = \frac{(26.5)(2.42)}{3(53)} \sqrt{(53^2 - 19^2)^3}$$

$$V = 48,851.88 \text{ in}^3 = 211.5 \text{ gal}$$

Elliptical Head in Vertical Position

The volume of the top portion, shown as section 2 of Figure 2.20, is

$$V_2 = \pi R_i^2 \left[y - \frac{y^3}{3(IDD)^2} \right] \tag{2.96}$$

The volume of the bottom portion, shown as section 1 in Figure 2.20, is

$$V_1 = \frac{2\pi(IDD)R_i^2}{3} - \pi R_i^2 \left[y - \frac{y^3}{3(IDD)^2} \right] \tag{2.97}$$

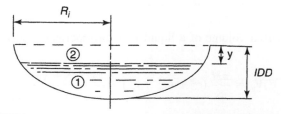

FIGURE 2.20 Partial volume of an elliptical head in the vertical position.

EXAMPLE 2.10

For the same head in the preceding example for the elliptical head, determine the partial volume for a vertical head with 19 inches of liquid. Using
Eq. 2.97, you get the following:

$$V = \frac{2\pi(26.50)(53.0)^2}{3} - \pi(53.0)^2 \left[7.5 - \frac{(7.5)^3}{3(26.50)^2} \right]$$

$$V = 155,903.62\,in^3 - 64,418.36\,in^3 = 91,485.26\,in^3$$

or, $V = 396.04$ gal

Partial Volumes of Torispherical Heads

The equations for a liquid occupying partial volumes of torispherical heads
follow.

Figures 2.21 and 2.22 use the following nomenclature:

V_k = knuckle volume
V_D = dish volume
KR = knuckle radius
y = height of liquid
IDD = inside depth of dish
ρ = inside dish radius

Torispherical Head in the Vertical Position

For the torispherical head in the vertical position shown in Figure 2.21(c), the
knuckle-cylinder partial volume is

$$V_k = \frac{\pi y}{6}(r_o^2 + 4r_m^2 + r_i^2) \qquad (2.98)$$

The partial volume of the dish region of a torispherical head in the vertical
position is

$$V_D = \frac{\pi y(3x^2 + y^2)}{6} \qquad (2.99)$$

The total partial volume of a liquid in a torispherical head in the vertical
position is

$$V_V = \frac{\pi H}{6}(r_o^2 + 4r_m^2 + r_i^2) + \frac{\pi y(3x^2 + y^2)}{6} \qquad (2.30)$$

where $y_d = IDD - KR$.

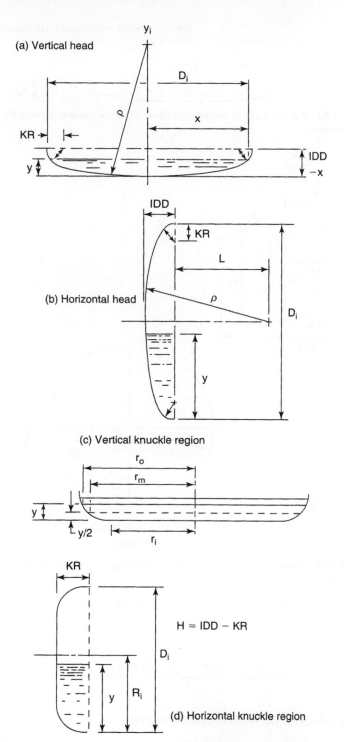

FIGURE 2.21 Partial volume of torispherical heads: **(a)** vertical position, **(b)** horizontal position, **(c)** knuckle region in vertical position, **(d)** knuckle region in horizontal position.

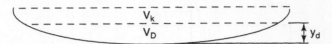

FIGURE 2.22 Partial volume of torispherical head in vertical position showing the dish and knuckle volumes.

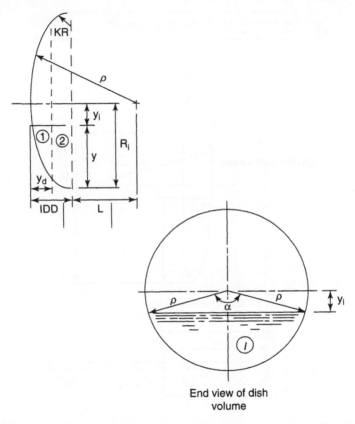

End view of dish
volume

FIGURE 2.23 Sketch for the example of the partial volume in a torispherical head in the horizontal position.

Torispherical Head in the Horizontal Position

In Figure 2.23 the partial volume of Dish 1 is

$$V_1 = \alpha \left| \frac{\sqrt{(\rho^2 - y_i^2)^3} - \sqrt{(\rho^2 - R_i^2)^3}}{3} - \frac{L(R_i^2 - y_i^2)}{2} \right| \qquad (2.101)$$

The volume of the knuckle-cylinder region is

$$V_2 = \alpha \left[\frac{4(KR)}{3\pi} + (R_i - KR) + (R_i - KR)^2 \right] \qquad (2.102)$$

The total partial volume for a torispherical head in the horizontal position is as follows:

$$V_T = V_1 + V_2 \tag{2.103}$$

$$V_T = \alpha \left| \frac{\sqrt{(\rho^2 - y_i^2)^3} - \sqrt{(\rho^2 - R_i^2)^3}}{3} - \frac{L(R_i^2 - y_i^2)}{2} \right|$$

$$+ \alpha \left[\frac{4(KR)}{3\pi} + (R_i - KR) + (R_i - KR)^2 \right] \tag{2.104}$$

where $L = \rho - IDD$.

EXAMPLE 2.11: A Torispherical Head in the Horizontal Position

A 102-inch OD F&D (flanged and dished, or torispherical) head made to ASME specifications ($KR \geq 0.60\rho$ and $KR > 3t_h$, where t_h = head thickness) is spun from a 1-inch plate. The head is in the horizontal position. The liquid level is 35 inches inside the head. Determine the volume of the liquid that occupies the partial volume of the head.

From the vessel head manufacturer's catalog and Figure 2.24, you can calculate the following:

$$\rho = 96 \, in, KR = 6.125 \, in, IDD = 17.562 \, in$$

$$R_i = \frac{100}{2} = 50 \, in, L = 96.0 - 17.562 = 78.438 \, in$$

From Eq. 2.104, you can determine

$$V_T = 2.532 \left| \frac{\sqrt{(96^2 - 15^2)^3} - \sqrt{(96^2 - 50^2)^3}}{3} - \frac{(78.438)(50^2 - 15^2)}{2} \right|$$

$$+ 2.532 \left[\frac{4(6.125)}{3\pi} + (50 - 6.125) + (50 - 6.125)^2 \right]$$

$$V_T = 34,093.44 \, in^3 = 147.59 \, gal$$

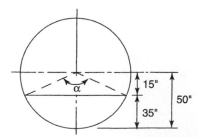

FIGURE 2.24   Example of torispherical head in the horizontal position.

EXAMPLE 2.12: A Torispherical Head in the Vertical Position

A 138-inch OD F&D head *not* made to ASME specifications is spun from a 1 ½-inch plate. The liquid level is 18 inches. Find the volume of the liquid.

From the vessel head manufacturer's catalog, you can calculate the following:

$$\rho = 132\,\text{in}, \text{KR} = 3\,\text{in}, \text{IDD} = 20.283\,\text{in}$$

$$R_i = \frac{138 - 2(1.5)}{2} = 67.50\,\text{in}$$

$$x = 67.50 - \left[3^2 - H^2\right]^{0.5} = 66.466\,\text{in}$$

For the knuckle-cylinder region,

$$r_o = R_i = 67.50; r_i \cong R_i - \text{KR} = 67.50 - 3.00 = 64.50\,\text{in}$$

$$r_m = \frac{67.50 + 64.50}{2} = 66.0$$

$$h = \left|20.283 - (3.0 + 15.0)\right| = 2.283\,\text{in}$$

$$V_V = \frac{\pi(2.283)}{6}\left[67.50^2 + 4(66.0)^2 + 64.5^2\right] + \frac{\pi(17.283)\left[3(64.5)^2 + 17.283^2\right]}{6}$$

$$V_V = 31,247.726\,\text{in}^3 + 115,645.832\,\text{in}^3$$

$$V_V = 146,893.558\,\text{in}^3 = 635.903\,\text{gal}$$

REFERENCES

1. ASME Section VIII, Division 1, Boiler and Pressure Vessel Code, 2007.
2. API Recommended Practice 579, *Fitness-for-Service*, 1st edition, American Petroleum Institute, January 2000.

Dynamic Response of Pressure Vessels and Stacks

This chapter describes the dynamic response of pressure vessel columns and stacks. We will focus on screening criteria and methods of remediation and follow with a discussion on methodology.

The subject of a vertical column or stack responding to wind in dynamic resonance has been addressed in engineering for over 100 years. For the past 50 years, discussion of the subject has grown in engineering publications. The subject is more accurately called *fluid-structure interaction*. The dynamic resonance is mostly caused by vortex shedding around the column or stack, but where there are two or more stacks, the mechanism of turbulence buffeting can exist.

It has been found that the upper third or fourth of the tower is significant because the correlation length of the vortices mostly affects this portion. The correlation length applies to the length over which the vortex streets are synchronized with each other. If two vortex streets around a tower are acting at different elevations but are in phase with each other, the distance between the two elevations of the vortices is called the *correlation length.* On a vertical tower, the correlation length is usually the upper third or fourth of the tower. Below this level the vortices diminish rapidly in magnitude. The correlation length can be two-dimensional in one plane or three-dimensional in three planes. In the former case, the tower behaves much like a pendulum. In the latter case, the tower top moves in an elliptical orbit in which the major axis of the ellipse is normal to the air flow. This latter case is the most common response. Readings from accelerometers mounted on the tops of towers reveal that the elliptical path is not a pure ellipse, but highly irregular jagged patterns that approximate an ellipse. Towers with two or more diameters with a significant amount of mass in the top one third or fourth tend to be problematic. The piping, platforms, and ladders can act as vortex inhibitors, but the designer should not depend on this outcome.

This elliptical pattern is what in the past was called *ovaling*. Some towers display an oval movement in resonance more than others; however, field accelerometers show that the movement may not be entirely elliptical; it is random in a generally elliptical pattern. Several screening criteria can be used to predict dynamic resonance response; these criteria are as follows:

1. Critical wind velocity is the wind speed in which dynamic resonance occurs, V_1. This is defined as follows:

$$V_1 = \frac{3.40d(\text{ft})}{T\left(\dfrac{\text{sec}}{\text{cycle}}\right)} \tag{3.1}$$

where
d = average diameter of the top third or fourth portion of a tower
T = period of vibration

This is the first critical wind velocity and is usually the one that governs compared to the second critical wind velocity, which is defined as follows:

$$V_2 = 8.25V_1 \tag{3.2}$$

If the critical wind velocity is close to the hourly averaged wind speed, resonance is possible. This means that if the prevailing wind remains constant over a prolonged time span, large dynamic amplitudes can be possible. Short wind gusts can set up resonance, but usually it is only temporary. When resonance occurs in the field, there is difficulty in measuring wind speed. This parameter requires judgment but can be helpful when you are using the other criteria.

2. The vortex shedding frequency is the frequency in which the vortices will shed. It is defined as follows:

$$f_v = \frac{0.2\tilde{v}\left(\dfrac{\text{ft}}{\text{sec}}\right)}{d(\text{ft})} = \text{Hertz} \tag{3.3}$$

where
$\tilde{v}$ = 45 mph or 66 fps
d = average diameter defined previously

It has been found in various studies that vortex shedding develops over the length of the top one third or top fourth of the tower, be it a process column or stack. If any top one third or fourth of the tower when $f_r < 2f_v$, then an oval pattern response is possible and likely. The wind velocity that would theoretically induce ovaling is

$$V_o = \frac{60f_r d}{2S} \tag{3.4}$$

where S = Strouhal number = 0.2 for most applications.

3. The computation of the natural frequency method using Rayleigh's method, which will be discussed later, can be applied to estimate the maximum dynamic displacement of the tower. This method is for undamped systems but is reasonably accurate in computing the first natural period, T, for most engineering applications. If you were to theoretically place the tower horizontally and fix it like a cantilever beam at the base, when it is subjected to gravity, the resulting deflection would be a reasonable estimate of the potential maximum response during resonance. This has been a helpful, and somewhat accurate, prediction of the dynamic response. The criterion of 6 inches per 100 feet, or 0.5% of the total height, is used to determine the section modulus of each tower section to minimize the response for process columns with trays. For packed columns, 9 inches per 100 feet, or 0.75% of the total height, is a criterion. Using this approach, you can change each tower section thickness to alter the mass distribution along the tower height. This technique has been a helpful rule of thumb in industrial practice; however, some members of academia do not like its use. This book is not intended for these people; the target audience here is for people working in the real world.

4. I developed the plot in Figure 3.1 [1] by comparing data for approximately 100 stacks. These data were taken from stacks, not process columns. The difference can be significant because the latter have more external attachments, such as piping, ladders, and platforms, and process fluids during operation. You must be cognizant of this difference when using the plot in Figure 3.1.

   The dynamic response of stacks largely depends on the structural damping coefficient, ζ. Table 3.1 provides a list of structural damping coefficients for stacks. Using this table, you can select a structural damping value and refer to Figure 3.1 to predict a dynamic displacement.

   As you can see from the curves, for one value of natural frequency, the lower the structural damping coefficient, the higher the ratio of the dynamic amplitude to the total height, Δ. Thus, for one natural frequency, it is possible to obtain a range of dynamic amplitudes between the upper and lower bound curves for the design case. If the natural frequency is lower than the value shown in the figure, then you have several options.

   If the situation is the design phase of constructing a stack then one can increase the natural frequency and/or increase the structural damping coefficient and/or add vibration inhibitors. The latter will be discussed under the topic of remediation. Also, the criterion for the minimum values of natural frequency is discussed later in the chapter. Each stack is influenced by its surroundings, including the support at the base. If the stack is mounted on another piece of equipment, then both the stack and supporting equipment need to be assessed together. Also, a stack downstream from other stacks will respond differently from one standing alone with no other stack in the proximity. This explains some of the wide scatter in the data.

   The vast majority of empirical data regarding vibration is widely scattered. It was found that for the same natural frequency and structural damping

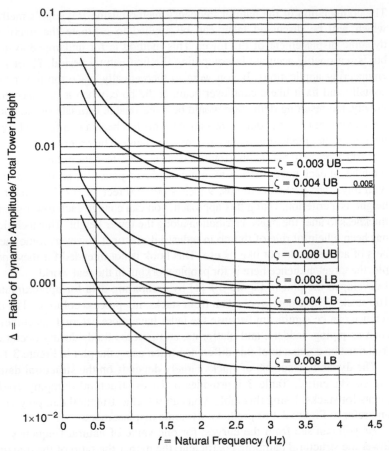

FIGURE 3.1   Probabilistic plot of the stack's natural frequency f versus Δ, the ratio of the maximum dynamic amplitude (in) to the total height (in) for various values of structuralt damping (ζ). UB = the upper bound value for Δ for a given frequency value; LB = the lower bound frequency value for Δ.

coefficient, various towers exhibited different dynamic amplitudes. The scatter is shown in Figure 3.1; the lower line shows the dynamic amplitude for a certain natural frequency and structural damping coefficient, while the upper curves do the same for the higher bound values. Figure 3.1 shows only the extreme values of ζ. You can interpolate for the value of ζ in the figure, using prudent judgment of the application at hand.

5. From steps 3 and 4 earlier in this section, you may suspect that another important parameter for screening dynamic response is the fundamental natural frequency of the tower. I observed that in process columns with the mass distributed toward the upper sections, there was a correlation between the fundamental natural frequency and excessive dynamic response. I observed

TABLE 3.1 Industrial Accepted Structural Damping Coefficient, ζ, Values
$\zeta_T = \zeta_{min} + \Sigma\zeta_i \leq 0.008$

Stacks Supported on Ground	ζ Values	Stacks Supported on Elevated Steel	ζ Values
Minimum Value— unlined stack, all welded, on rock, or very stiff soil	$\zeta_{min} = 0.004$	Minimum Value— unlined stack, all welded, on steel (such as a furnace)	$\zeta_{min} = 0.003$
ζ Values added to ζ_{min}	ζ_I Values	ζ Values added to ζ_{min}	ζ_I Values
• Stack lining (min. 2" thick)	0.002	• Stack lining (min. 2" thick)	0.001–0.002 (1)
• Stack constructed with a minimum of 5 flanges	0.002	• Stack constructed with a min. of 5 flanges	0.001–0.002 (1)
• External piping attaching from 60°–120° between each other	0.003	• External piping attaching from 60°–120° between each other	0.003
• Stack mounted on soft soil	0.001	• Steel support provided with refractory casing	0.002
$\zeta_T > 0.008$ if:		$\zeta_T > 0.008$ if:	
• 3 guy wires attached to upper ¼ or ⅓ of height	$\zeta_T = 0.012$	• 3 guy wires attached to upper ¼ or ⅓ of height	$\zeta_T = 0.012$
• Installation of proper damping pads	$\zeta_T = 0.012$	• Installation of proper damping pads	$\zeta_T = 0.012$

Note: A higher damping value is used if a relatively flexible stack is mounted on a stiff strcuture (Structure stiffness > 100x stack stiffness)

that the fundamental frequency of the tower should be greater than 1.0 Hz, although slightly less than 1.0 Hz would be acceptable. If the natural frequency is not below 0.97 Hz, excessive dynamic amplitude will be avoided. This test has been applied to many process columns, and towers with a natural frequency less than 0.97 Hz failed. **Note:** This criterion does *not* say that, if the natural frequency of the tower in the first mode is greater than 0.97 Hz, then dynamic resonance will not occur.

6. The ASME STS-1 Steel Stacks [2] governing standard used in the design of steel stacks is important, and everyone should be familiar with it. In Example E.7 in the standard, the mass damping parameter is used as a

criterion for assessing dynamic response. This parameter is developed as follows:

$$m_r = \frac{m_e}{\rho \bar{D}^2 g} \tag{3.5}$$

where

m_r = dimensionless mass

ρ = density of air, taken in the example as 0.00238 slugs/ft^3

$\bar{D}$ = average diameter of the stack, ft

g = 32.17 lb$_m$/slug

Note: If the value of the density is entered as lb$_m$/ft^3, then

$$\rho^e = 0.00238 \left(\frac{\text{slugs}}{\text{ft}^3} \right) g(32.17) \frac{\text{lb}_m}{\text{slug}} = 0.0779 \frac{\text{lb}_m}{\text{ft}^3}$$

and Eq. 3.5 becomes

$$m_r = \frac{m_e}{\rho^e \bar{D}^2} \tag{3.5a}$$

The structural damping value, ξ_s (β_s is used in the standard), is given in Table 3.2 and is taken from Table 5.2.1 in the standard [2].

The mass damping parameter is defined as follows:

$$m_p \equiv m_r \zeta_s \tag{3.6}$$

When referring to stacks, you can find the structural damping value, ξ_s, in Table 3.2. When referring to process columns, you use either Table 3.1 or 3.2.

Example E.7 in ASME STS-1 [2] says that $m_p > 0.8$ for the *stack* to be satisfactory. Findlay [3] reports that ExxonMobil requires $m_p > 1.1$ for the

---

TABLE 3.2 ASME STS-1 Representative Structural Damping Values (ξ_s)

Support	Damping Value	
Type Welded Stack	Rigid Support (1)	Elastic Support (2)
Unlined	0.002	0.004
Lined (3)	0.003	0.006

NOTES: (1) *Foundations on bedrock, end-bearing piles or other rigid base support conditions.*
(2) *For foundations with friction piles or mat foundations on soil or other elastic base support conditions.*
(3) *Lining must consist of a minimum 2 in. thick, nominally 100 pcf density liner material for stack to be considered lined for the use of this table. (Reprinted from ASME STS-1-2006, by permission of the American Society of Mechanical Engineers. All rights reserved.)*

stack or *process column* to be satisfactory to prevent unacceptable dynamic amplitudes.

We will illustrate the six fluid-structure interaction criteria in Example 3.1. It is important that you are cognizant of the fact that these criteria are based on rules of thumb. They are not to be considered as laws of physics, and judgment has to be rendered in their application.

Example 3.1

This example (from [1]) was an actual case where a tower developed dynamic amplitudes that were unacceptable. The internals, ladders, and external piping were installed on the process column. We will evaluate each of the six criteria and see how they apply to this process column, which is shown in Figure 3.2.

First, you need to find the natural frequency. To this end, refer to Figure 3.2 and Figure 3.3 for the solution.

After solving for the natural frequency, you can continue with the following calculations.

The first critical wind velocity is

$$V_1 = \frac{3.40d}{T}$$

$$L = \frac{76.96}{4} = 19.24\,\text{ft}$$

$$d = \left(\frac{13.0}{19.24}\right)(8.74) + \left(\frac{6.24}{19.24}\right)(3.75) = 7.122$$

From Eq. 3.1, making the vortex shedding frequency equal to the natural frequency, you can calculate

$$V_1 = \frac{f_v d}{s} = \frac{(0.97)(7.122)}{0.2} = 34.540\,\frac{\text{ft}}{\text{sec}}$$

Considering the top portion (Section 1), you can find

$$V_1 = \frac{(0.97)(8.74)}{0.2} = 42.39\,\frac{\text{ft}}{\text{sec}} = 28.90\,\text{mph}$$

Since the field measurements indicated an air velocity of 30 mph and a column dynamic amplitude of 13 inches, this agrees well with the previous calculations with a possible amplitude of 13.59 inches. For a stack only 77 feet 5 inches tall, the 13.59 inches is significantly higher than the 6 inches per 100 feet criterion. Thus, if you use criteria 1, 2, and 5, described earlier in the chapter, the process column is a potential vibration problem.

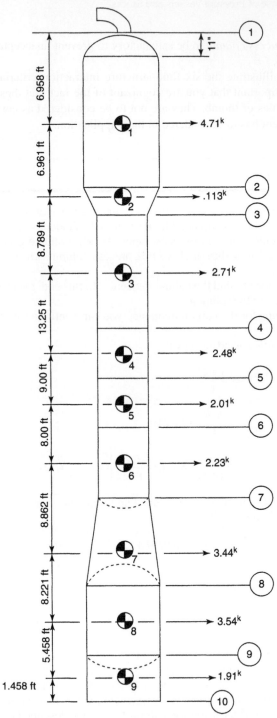

FIGURE 3.2 Schematic showing the process column that experienced unacceptable dynamic amplitudes.

EXAMPLE OF A SOLUTION FOR NATURAL FREQUENCY

Rules Sheet

Rules

M1 = 0

M2 = M1 + W1 · L2

M3 = M2 + (W1 + W2) · L3

M4 = M3 + (W1 + W2 + W3) · L4

M5 = M4 + (W1 + W2 + W3 + W4) · L5

M6 = M5 + (W1 + W2 + W3 + W4 + W5) · L6

M7 = M6 + (W1 + W2 + W3 + W4 + W5 + W6) · L7

M8 = M7 + (W1 + W2 + W3 + W4 + W5 + W6 + W7) · L8

M9 = M8 + (W1 + W2 + W3 + W4 + W5 + W6 + W7 + W8) · L9

M10 = M9 + (W1 + W2 + W3 + W4 + W5 + W6 + W7 + W8 + W9) · L10

$$RAT2 = \frac{M2 \cdot 1000}{I2}$$

$$RAT3 = \frac{M3 \cdot 1000}{I3}$$

$$RAT4 = \frac{M4 \cdot 1000}{I4}$$

$$RAT5 = \frac{M5 \cdot 1000}{I5}$$

$$RAT6 = \frac{M6 \cdot 1000}{I6}$$

$$RAT7 = \frac{M7 \cdot 1000}{I7}$$

$$RAT8 = \frac{M8 \cdot 1000}{I8}$$

$$RAT9 = \frac{M9 \cdot 1000}{I9}$$

$$RAT10 = \frac{M10 \cdot 1000}{I10}$$

$$S10 = \frac{(RAT10 + RAT9) \cdot L10}{2}$$

$$S9 = \frac{(RAT9 + RAT8) \cdot L9}{2}$$

$$S8 = \frac{(RAT8 + RAT7) \cdot L8}{2}$$

$$S7 = \frac{(RAT7 + RAT6) \cdot L7}{2}$$

FIGURE 3.3(a)   Equation sheet for solution for natural frequency.

$$S6 = \frac{(RAT6 + RAT5) \cdot L6}{2}$$

$$S5 = \frac{(RAT5 + RAT4) \cdot L5}{2}$$

$$S4 = \frac{(RAT4 + RAT3) \cdot L4}{2}$$

$$S3 = \frac{(RAT3 + RAT2) \cdot L3}{2}$$

$$S2 = \frac{RAT2 \cdot L2}{2}$$

$\Sigma S10 = S10$

$\Sigma S9 = S10 + S9$

$\Sigma S8 = \Sigma S9 + S8$

$\Sigma S7 = \Sigma S8 + S7$

$\Sigma S6 = \Sigma S7 + S6$

$\Sigma S5 = \Sigma S6 + S5$

$\Sigma S4 = \Sigma S5 + S4$

$\Sigma S3 = \Sigma S4 + S3$

$\Sigma S2 = \Sigma S3 + S2$

$$P10 = \frac{\Sigma S10 \cdot L10}{2}$$

$$P9 = \frac{(\Sigma S10 + \Sigma S9) \cdot L9}{2}$$

$$P8 = \frac{(\Sigma S9 + \Sigma S8) \cdot L8}{2}$$

$$P7 = \frac{(\Sigma S8 + \Sigma S7) \cdot L7}{2}$$

$$P6 = \frac{(\Sigma S7 + \Sigma S6) \cdot L6}{2}$$

$$P5 = \frac{(\Sigma S6 + \Sigma S5) \cdot L5}{2}$$

$$P4 = \frac{(\Sigma S5 + \Sigma S4) \cdot L4}{2}$$

$$P3 = \frac{(\Sigma S4 + \Sigma S3) \cdot L3}{2}$$

$$P2 = \frac{(\Sigma S3 + \Sigma S2) \cdot L2}{2}$$

$F10 = P10$

$F9 = P10 + P9$

$F8 = F9 + P8$

$F7 = F8 + P7$

FIGURE 3.3(a) (Continued)

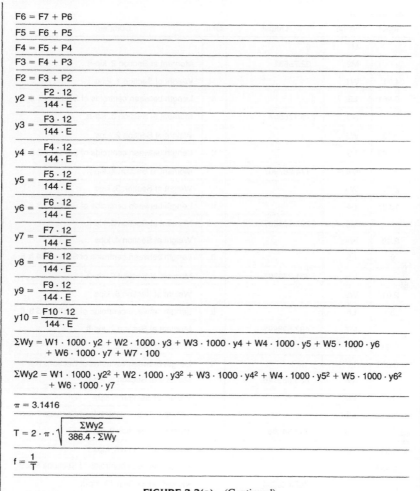

$F6 = F7 + P6$

$F5 = F6 + P5$

$F4 = F5 + P4$

$F3 = F4 + P3$

$F2 = F3 + P2$

$$y2 = \frac{F2 \cdot 12}{144 \cdot E}$$

$$y3 = \frac{F3 \cdot 12}{144 \cdot E}$$

$$y4 = \frac{F4 \cdot 12}{144 \cdot E}$$

$$y5 = \frac{F5 \cdot 12}{144 \cdot E}$$

$$y6 = \frac{F6 \cdot 12}{144 \cdot E}$$

$$y7 = \frac{F7 \cdot 12}{144 \cdot E}$$

$$y8 = \frac{F8 \cdot 12}{144 \cdot E}$$

$$y9 = \frac{F9 \cdot 12}{144 \cdot E}$$

$$y10 = \frac{F10 \cdot 12}{144 \cdot E}$$

$\Sigma Wy = W1 \cdot 1000 \cdot y2 + W2 \cdot 1000 \cdot y3 + W3 \cdot 1000 \cdot y4 + W4 \cdot 1000 \cdot y5 + W5 \cdot 1000 \cdot y6 + W6 \cdot 1000 \cdot y7 + W7 \cdot 100$

$\Sigma Wy2 = W1 \cdot 1000 \cdot y2^2 + W2 \cdot 1000 \cdot y3^2 + W3 \cdot 1000 \cdot y4^2 + W4 \cdot 1000 \cdot y5^2 + W5 \cdot 1000 \cdot y6^2 + W6 \cdot 1000 \cdot y7$

$\pi = 3.1416$

$$T = 2 \cdot \pi \cdot \sqrt{\frac{\Sigma Wy2}{386.4 \cdot \Sigma Wy}}$$

$$f = \frac{1}{T}$$

FIGURE 3.3(a)   (Continued)

Variables sheet

Input	Name	Output	Unit	Comment
	M1	0		
	M2	32.78631		Moment at Section 2, kip-ft
4.71	W1			Weight of Section 1, kips
6.961	L2		ft	Length between centroids of Sections 1 & 2
	M3	75.175657		Moment at Section 3, kip-ft
.113	W2			Weight at Section 2, kips
8.789	L3		ft	Length between centroids of Sections 2 & 3
	M4	174.987907		Moment at Section 4, kip-ft
2.71	W3			Weight at Section 3, kips
13.25	L4		ft	Length between centroids of Sections 3 & 4
	M5	265.104907		Moment at Section 5, kip-ft
2.48	W4			Weight at Section 4, kips
9	L5		ft	Length between centroids of Sections 4 & 5
	M6	361.288907		Moment at Section 6, kip-ft
2.01	W5			Weight at Section 5, kips
8	L6		ft	Length between centroids of Sections 5 & 6
	M7	487.598993		Moment at Section 7, kip-ft
2.23	W6			Weight at Section 6, kips
8.862	L7		ft	Length between centroids of Sections 6 & 7
	M8	633.053146		Moment at Section 8, kip-ft
3.44	W7			Weight at Section 7, kips
8.221	L8		ft	Length between centroids of Sections 7 & 8
	M9	748.94286		Moment at Section 9, kip-ft
3.54	W8			Weight at Section 8, kips
5.458	L9		ft	Length between centroids of Sections 8 & 9
	M10	782.685354		Moment at Section 10, kip-ft
1.91	W9			Weight at Section 9, kips
1.458	L10		ft	Length between centroids of Sections 9 &10
	RAT2	117513.655914		Ratio of moment to moment of inertia
.279	I2			Moment of inertia of Section 2, ft^4
	RAT3	1105524.367647		Ratio of moment to moment of inertia
.068	I3			Moment of inertia of Section 3, ft^4
	RAT4	1698911.718447		Ratio of moment to moment of inertia
.103	I4			Moment of inertia of Section 4, ft^4
	RAT5	1907229.546763		Ratio of moment to moment of inertia
.139	I5			Moment of inertia of Section 5, ft^4

FIGURE 3.3(b) Variable sheet showing results and answers for solution of natural frequency.

Variables sheet

Input	Name	Output	Unit	Comment
	RAT6	2041180.265537		Ratio of moment to moment of inertia
.177	I6			Moment of inertia of Section 6, ft^4
	RAT7	3750761.484615		Ratio of moment to moment of inertia
.13	I7			Moment of inertia of Section 7, ft^4
	RAT8	1120448.046018		Ratio of moment to moment of inertia
.565	I8			Moment of inertia of Section 8, ft^4
	RAT9	1872357.150000		Ratio of moment to moment of inertia
.4	I9			Moment of inertia of Section 9, ft^4
	RAT10	1956713.385000		Ratio of moment to moment of inertia
.4	I10			Moment of inertia of Section 10, ft^4
	S10	2791392.420015		Intermediate moment ratio
	S9	8167365.379932		Intermediate moment ratio
	S8	20023106.775667		Intermediate moment ratio
	S7	25664093.894924		Intermediate moment ratio
	S6	15793639.249197		Intermediate moment ratio
	S5	16227635.693441		Intermediate moment ratio
	S4	18579389.070371		Intermediate moment ratio
	S3	5374640.594539		Intermediate moment ratio
	S2	409006.279409		Intermediate moment ratio
	ΣS10	2791392.420015		Sum of intermediate ratio
	ΣS9	10958757.799947		Sum of intermediate ratio
	ΣS8	30981864.575615		Sum of intermediate ratio
	ΣS7	56645958.470539		Sum of intermediate ratio
	ΣS6	72439597.719736		Sum of intermediate ratio
	ΣS5	88667233.413177		Sum of intermediate ratio
	ΣS4	107246622.483548		Sum of intermediate ratio
	ΣS3	112621263.078087		Sum of intermediate ratio
	ΣS2	113030269.357495		Sum of intermediate ratio
	P10	2034925.074191		Force per unit length
	P9	37524159.950277		Force per unit length
	P8	172396928.274747		Force per unit length
	P7	388278883.917505		Force per unit length
	P6	516342224.761098		Force per unit length
	P5	724980740.098109		Force per unit length
	P4	1297929295.315800		Force per unit length
	P3	966209423.100602		Force per unit length

FIGURE 3.3(b)   (Continued)

Variables sheet

Input	Name	Output	Unit	Comment
	P2	785380158.642043		Force per unit length
	F10	2034925.074191		Section force per unit length
	F9	39559085.024468		Section force per unit length
	F8	211956013.299215		Section force per unit length
	F7	600234897.216720		Section force per unit length
	F6	1116577121.977820		Section force per unit length
	F5	1841557862.075930		Section force per unit length
	F4	3139487157.391730		Section force per unit length
	F3	4105696580.492330		Section force per unit length
	F2	4891076739.134370		Section force per unit length
	y2	13.586324		Deflection of vessel section
3E7	E			Modulus of elasticity of metal in vessel section, psi
	y3	11.404713		Deflection of vessel section
	y4	8.720798		Deflection of vessel section
	y5	5.115439		Deflection of vessel section
	y6	3.101603		Deflection of vessel section
	y7	1.667319		Deflection of vessel section
	y8	0.588767		Deflection of vessel section
	y9	0.109886		Deflection of vessel section
	y10	0.005653		Deflection of vessel section
	ΣWy	113977.464553		
	ΣWy2	1181876.417781		
	π	3.1416		
	T	1.029293		First period of vibration, sec/cycle
	f	0.971541		Natural frequency of first mode, Hz

FIGURE 3.3(b)   (Continued)

Now look at the sixth criterion: the mass damping parameter. According to Table 3.2 (from Table 5.2.1 of ASME STS-1 [2]), an unlined stack has a structural damping value, x, of 0.004, which Findlay [3] states is commonly used. According to Note 3 of this table, a structural damping value of 0.006 can be used only when the stack has a lining of a minimum of 2 inches thick and a nominal density of 100 pounds per cubic foot (pcf). A density of 100 pcf is very dense—like fire proofing. It can be argued that, for an operating process column with an overhead pipe extending most of the full length and with ladders and internal attachments, you can use a structural damping coefficient of 0.006. Also ASME STS-1 [2] is for stacks, not process columns. Now see what the sixth criterion yields.

Computing the average diameter for the top third of the column, you learn that

$$L_{T\frac{1}{3}} = \frac{77.417}{3} = 25.806 \text{ ft}$$

$$\bar{D} = \left(\frac{14.417}{25.806}\right) 36'' + \left(\frac{11.389}{25.806}\right) 24'' = 30.704''$$

or $d = 2.559$ ft.

For the top head, and referring to Chapter 2 for the weight of a 36-inch ID 2:1 ellipsoidal head with a minimum 5/16-inch head, you can calculate

$$BD = 1.22(ID) + 2(S.F.) + T \qquad\qquad (2.82)$$

$$BD = 1.22(36) + 2(2) + \frac{5''}{16} = 48.23''$$

$$\text{WgtHD} = (0.283)\frac{\text{lb}_m}{\text{in}^3}\left(\frac{\pi}{4}\right)(48.23)^2\,\text{in}^2\left(\frac{5}{16}\right)\text{in} = 161.571\,b_m$$

For the 36-inch ID cylindrical can, the weight is

$$\text{Wgt36} = \left(\frac{\pi}{4}\right)(36.625^2 - 36^2)\,\text{in}^2(162)\,\text{in}\,(0.283)\frac{\text{lb}_m}{\text{in}^3} = 1634.4\,\text{lb}_m$$

For the 24-inch cylindrical can, the weight is

$$L = 25.806 \text{ ft} - \left(\frac{162 + 9}{12}\right) = 11.556 \text{ ft} = 138.672 \text{ in}$$

$$\text{Wgt24} = \left(\frac{\pi}{4}\right)(24.5^2 - 24^2)\,\text{in}^2(138.672)\,\text{in} = 747.44\,\text{lb}_m$$

For the top third of the stack, the weight is

$$Wgt = 161.57 + 1634.4 + 747.44 = 2543.41 \, lb_m$$

The mass per unit length becomes

$$m_e = \frac{2543.41 \, lb_m}{25.806 \, ft} = 98.56 \, \frac{lb_m}{ft}$$

Now using a structural damping value of $\zeta_s = 0.004$ and an air density of $0.0779 \, lb_m/ft^3$, you can substitute into Eq. 3.6 to obtain

$$m_p = m_r \zeta_s = \frac{m_e \zeta_s}{\rho \bar{D}} = \frac{(98.56) \, \dfrac{lb_m}{ft} \, (0.004)}{(0.0779) \, \dfrac{lb_m}{ft^3} \, (2.559)^2 ft^2} = 0.773 < 0.8$$

Using the structural damping value of $\zeta_s = 0.006$, you obtain

$$m_p = \frac{0.006}{0.004}(0.773) = 1.1592 > 1.1$$

Depending on which structural damping value you use, the sixth criterion becomes less certain, and opinions run high on each side for each value of ζ_s. Hence, it is recommended that all six criteria be used together to provide closure to the problem. If one of the criteria indicates the tower is unacceptable, then this makes the decision easier. Criterion 6 was intended for stacks, not process stacks, but like the other criteria, it can be handy.

FLOW-INDUCED VIBRATION-IMPEDING DEVICES

Helical Strakes

Devices can be built into stacks to counter the vortex shedding, which causes dynamic instability. Helical vortex strakes are the most common and practical vibration inhibitors for stacks. They are generally too awkward to use on process columns because of external attachments, such as ladders, platforms, and piping.

The application of helical vortex strakes to vertical cylindrical towers has shown remarkable results. I independently developed the method presented here over 20 years ago in a fabrication shop in Houston, Texas. Others developed the strake concept long before that, but the challenge for me at the time was to figure out how to fabricate and install them onto stacks using simple shop tools. This information was published first in a technical journal and then later in [1].

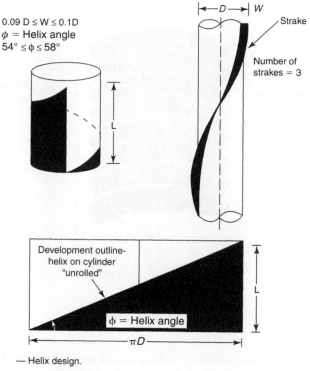

$0.09\,D \le W \le 0.1D$
ϕ = Helix angle
$54° \le \phi \le 58°$

Strake

Number of
strakes = 3

Development outline-
helix on cylinder
"unrolled"

ϕ = Helix angle

— Helix design.

FIGURE 3.4   Cylindrical strake helix geometry.

The strakes' function is to break up vortices such that mode shapes stimulating dynamic response to the tower are quickly dampened. It is significant to note that adding the strakes significantly increases the drag and thus wind loading. These strakes are shown in Figure 3.4.

To minimize the flow-induced drag and optimize the vortex-breaking effect, you should leave the strake width, W (ft), in the following range:

$$0.09D \le W \le 0.10D$$

where D = OD of stack, ft.

Figure 3.4 shows a helix generated on a cylinder by taking a template πD long by L high and wrapping it around a cylinder. The length, L, of the helix is the top third of the stack. Wind tunnel tests have shown that vortex-breaking devices are most effective on the upper third of the stack. The helix angle, ϕ, should have a magnitude in the following range:

$$54° \le \phi \le 58°$$

There are always three strakes per stack to counter the alternate formation of vortices on either side of the stack.

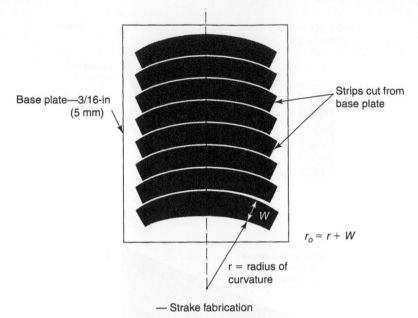

Base plate—3/16-in (5 mm)

Strips cut from base plate

W

$r_o = r + W$

r = radius of curvature

— Strake fabrication

FIGURE 3.5   Strake fabrication detail.

Strakes can be fabricated from a flat piece of metal, normally 3/16 inch or 5 mm thick. Each strake is divided to a certain number of strips, usually 5–20 segments, depending on the length of the stack. The overall length of the individual strakes that are divided is determined by

$$S = \sqrt{(\pi D)^2 + L^2}$$ (3.7)

where
D = the OD of the stack, as defined previously
L = height of the tower portion straked (one third of total stack height), ft

The number S is divided into individual strips that are cut from a larger piece of plate, as shown in Figure 3.5.

The strips must be cut to a radius of curvature, r, which is determined as follows:

$$r = \frac{a^2 \omega^2 + b^2}{a \omega^2}$$ (3.8)

where

$$a = \frac{D}{2}, \text{ ft}$$

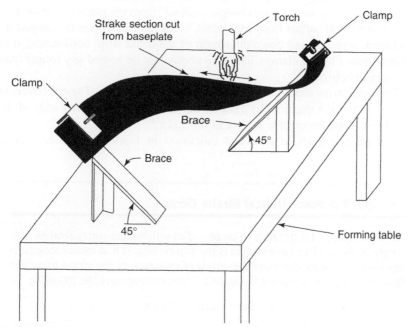

FIGURE 3.6 Clamping each strip on 45° offsets and hot-forming with a torch to obtain the desired geometry.

$$b = \frac{L}{2\pi\omega}$$

ω = number of revolutions around stack cylinder made by helical strake (usually $\omega = 1$).

An alternative formula which is within 2%–3% in error of Eq. 3.8, is

$$r = \frac{\lambda W}{1 - \lambda} \qquad (3.9)$$

where

$$\lambda = \frac{S_i}{S_o} = \frac{\text{interior arc length of helix}}{\text{exterior arc length of helix}} \qquad (3.10)$$

The value S_i is determined by using the outside diameter of the stack in Eq. 3.7, and S_o is obtained by using $D + 2W$ in place of the same equation. For the most accurate results, you should use Eq. 3.8 because it is the exact radius of curvature of a helix projected on a cylinder [4].

Strips are laid out, as shown in Figure 3.5, with an inner radius of curvature determined by Eq. 3.8 and an outer radius of $r = r + W$. You want the helix to be perpendicular to the centerline of the cylinder along the entire length of the helical strake shown in Figure 3.4. To obtain this, you place each metal strip in a rig, as shown in Figure 3.6.

The rig is composed of two clamps, each 45° from the plane perpendicular to the table, or 90° offset from each other. Once the metal strip is clamped in, a hot torch is run up and down the length of the metal strip, hot-forming it to a shape formed by the clamps. The strip should not be heated any longer than necessary to hot-form it.

The metal strips should be the same material as the stack. The effectiveness of the system is not impaired by a gap of 0.005D between the inner edge of the helical strake and the outside surface of the stack.

This method leads to ease and quickness in fabricating helical vortex strakes.

Example 3.2: Stack Helical Strake Design

An exhaust stack 126 ft. tall is to be provided with helical vortex strakes. The length of the stack to have strakes is the top portion 31 ft. 6 inches long (top fourth of the stack). Compute the radius of curvature of the strake to be cut from the flat plate. Referring to Figure 3.7, you can compute the following:

$$D = OD \text{ of stack} = 7 \text{ ft } 4 \text{ in}$$

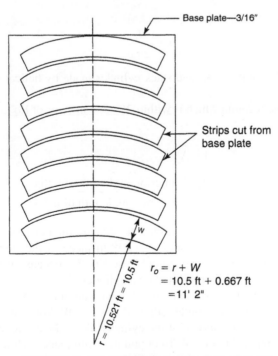

$$
\begin{aligned}
r_o &= r + W \\
&= 10.5 \text{ ft} + 0.667 \text{ ft} \\
&= 11' \, 2''
\end{aligned}
$$

FIGURE 3.7 Fabricating helical strakes from a flat plate.

$$L = 31\,ft\,6\,in$$

$$a = \frac{D}{2} = \frac{7.333}{2} = 3.667$$

$$\omega = 1$$

$$b = \frac{L}{2\pi\omega} = \frac{31.5}{2\pi(1)} = 5.013$$

$$r = \frac{a^2\omega^2 + b^2}{a\omega^2}$$

$$r = \frac{(3.667)^2(1)^2 + (5.013)^2}{(3.667)(1)^2} = 10.521\,ft$$

Check:

Using the alternate equation, you can calculate

$$S_i = \text{interior arc length} = \left[(\pi D_o)^2 + L^2\right]^{0.5} = 41.637\,ft$$

$$S_o = \text{exterior arc length} = \left[\left[\pi(8.667)\right]^2 + (31.5)^2\right]^{0.5} = 41.637\,ft$$

$$\lambda = \frac{S_i}{S_o} = \frac{39.025}{41.637} = 0.937$$

$$r = \frac{\lambda W}{1 - \lambda}$$

$$r\frac{(0.937)(0.667)}{1 - 0.937} = 9.966\,ft = 9\,ft\,11.594\,in$$

$$\% \text{ error} = \left(\frac{10.521 - 9.966}{10.521}\right)(100) = 5.276\% \text{ error}$$

The final product is shown in Figure 3.8. The actual value of r used on the flat plate cutout shown in Figure 3.7 was 10.521 ft.

(a)

(b)

FIGURE 3.8 Helical strakes fabricated using the method described in Example 3.2.

Damping Pads

The structural damping of a tower can be increased by the application of a damping pad, as shown in Figure 3.9. It consist of two components: one is a damping washer or pad between the anchor bolt and the chair plate; and the other is a large pad that fits between the base plate and leveling, or shim, plate on top of the concrete foundation. There are several kinds of damping pads, each with its own damping characteristics. Normally, a damping pad consists of three layers: a top and bottom elastomer layer with a cork layer in the middle. Common elastomers used are silicone or nitrate, and normally are sold under trade names. Several companies make these pads; one is Fabreeka International in Boston, Massachusetts, and another is Tech Products Corporation (TRC) out of Dayton, Ohio.

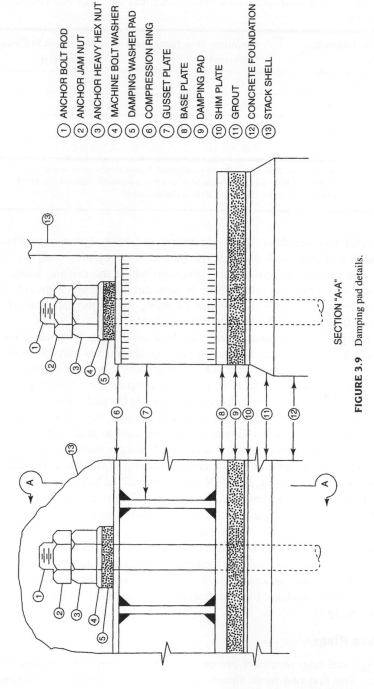

1. ANCHOR BOLT ROD
2. ANCHOR JAM NUT
3. ANCHOR HEAVY HEX NUT
4. MACHINE BOLT WASHER
5. DAMPING WASHER PAD
6. COMPRESSION RING
7. GUSSET PLATE
8. BASE PLATE
9. DAMPING PAD
10. SHIM PLATE
11. GROUT
12. CONCRETE FOUNDATION
13. STACK SHELL

SECTION "A-A"

FIGURE 3.9 Damping pad details.

TABLE 3.3 Minimum Structural Plate Thickness and Maximum Stiffener Spacing

Inside Diameter, D (ft)	Minimum Structural Plate Thickness*	Maximum Stiffener Spacing, ft
D ≤ 3.5	0.125	5 D
3.5 < D ≤ 6.5	0.1875	3 D
6.5 < D ≤ 18.0	0.1875	2 D
D ≥ 18.0	0.25	1½ D

*Note: Minimum plate thickness does not include corrosion allowance. If corrosion allowance is required, the minimum plate thickness will be increased by the magnitude of the corrosion allowance. (Reprinted from ASME STS-1-2006, by permission of the American Society of Mechanical Engineers.)

Several facilities have stacks exhibiting excessive dynamic amplitudes; the addition of these pads does not always add to the existing damping of the stack. The pads are effective "win" situations where the existing damping is very low, resulting in a net increase in structural damping. Although some manufacturers of the pads claim higher damping values, the maximum should be only 2.0% of critical damping, as listed in Table 3.3.

These pads are inexpensive and are best used for a new stack. For an existing operational stack, installation costs of these pads may be excessive. The disadvantages of these pads are (1) they cannot withstand temperatures >200°F and must be insulated from higher temperatures; (2) the pads may deteriorate when exposed to hydrocarbons; and (3) the pads may require maintenance and inspection. If the pads have deteriorated, their replacement may radically disrupt operations. Gunite linings are used to protect the pads from excessive temperatures or acids. These disadvantages must be considered at each site. If the increase in the structural damping of the stack still does not suffice for vortex shedding, then FIV inhibitors such as helical strakes or ovaling rings should be considered. We will cover ovaling rings next.

These damping pads are used for stacks mounted on concrete foundations. According to ASME STS-1, Paragraph 5.2.1(a) [2], for steel stacks mounted on buildings, the interaction effects of the building need to be included. Generally, this involves modeling the building and stack together. When stacks are mounted on steel structures, such as exhaust stacks, the steel structure and stack should be modeled together because the structure's stiffness will enter into the analysis.

Ovaling Rings

Placing metal rings on a stack can prevent dynamic amplitudes at higher mode shapes. The first two mode shapes are the stack in translation, retaining its circular shape. The third and fourth mode shapes involve the transformation of the circular outline of the stack cross-section to an elliptical shape. At the fifth and sixth mode shapes, the stack cross-section evolves into a triangular shape. These last two mode shapes do not occur with every stack because they are a function of the stack's diameter-to-thickness ratio (D/t). The occurrence of the higher mode shapes is known as *ovaling*. The ovaling rings provide a redistribution of the mass of the stack, which can offset the higher mode shapes. This can happen if the ovaling rings are located properly. To analyze the effect of the rings, you must consider the mass and moment of inertia of the ring as a separate section in Rayleigh's method, as shown in Example 3-1. The finite element method allows you to quickly see mode shapes and optimize the ring locations.

The phenomenon of ovaling is predominant with stacks. ASME STS-1, Paragraph 5.2.2(b) [2], gives guidelines for ovaling. For unlined steel stacks, the ovaling natural frequency is computed as follows:

$$f_o = \frac{680t}{D^2} \qquad (3.11)$$

where ASME STS-1 [2] defines
t = stack shell or liner wall thickness, inches
D = diameter of stack at elevation under consideration

The critical wind velocity for ovaling is

$$v_{co} = \frac{f_o D}{2S} \qquad (3.12)$$

where S = Strouhal number, mentioned previously in Eq. 3.4, usually taken as 0.2 for single stacks and may vary due to Reynolds numbers and multiple stacks.

If v_{co} is less than $\bar{V}_z$, the mean hourly wind speed (ft/sec), the unlined stack should be reinforced with ring stiffeners meeting the requirements of Table 3.3 (Table 4.4.7 in ASME STS-1 [2]).

The required minimum section modulus of the ring stiffener, S_s (in³), with respect to the neutral axis of its cross-section parallel to the longitudinal axis of the stack, is as follows:

$$S_s = \frac{(2.52 \times 10^{-3})(v_{co})^2 D^2 l_s}{\sigma_a} \qquad (3.13)$$

where
l_s = spacing between circumferential stiffeners, determined as the sum of half the distance to adjacent stiffeners on either side of the stiffener under consideration (ft)

$\sigma_a = 0.6F_y$, where F_y = specified minimum yield strength of the ring material at mean shell temperature, psi

In the area where helical strakes are attached to the stack, ring stiffeners may be omitted if you can prove that the helical strakes provide adequate stiffness.

Ovaling rings are used often as remediation devices. They can be welded in during shutdowns or turnarounds. You can avoid them in the design phase if you use thicker plates along the length of the stack using the criteria mentioned previously.

GUY CABLES, REMEDIATION DEVICES, AND SUPPORT OF FLARE STACKS

Guy cables can be remediation devices to stabilize a dynamic unstable stack or process column, or as a built-in design support mechanism for a tall and narrow diameter stack. A flare stack is more involved than other types of stacks because there are thermal gradients to complicate matters.

Determining the pre-tension required for guy cables on flare stacks is necessary for proper support. The method discussed here has been used in practice and is therefore tried and proven.

Flare stacks supported by guy cables are ubiquitous in use and seem quite innocuous to the untrained eye. However, their failure during the flaring of a stack could have dire consequences. This section is based on actual pretensioning cases that have proven to be successful.

This section concerns the static problem of guy wires. Experience has shown that if guy cables are properly tensioned, flow-induced vibration (FIV), induced by vortex shedding, is of minimal concern. The reason is that the natural frequency of the stack is well above the resonant range. One reason for pre-tensioning the cables is to avoid FIV.

Wind loadings should be checked with the appropriate wind standard or company standard. This section is based on a paper I [5] wrote while working in Saudi Arabia. The company in Saudi Arabia checked the wind loads at 100 mph, and the guy wires proved to be more than adequate.

The Basic Methodology

Flare stacks undergo several thermal regimes, hot and cold. If a cold liquid is flared, and on entering the flare stack, the burner tip does not ignite, then the entire stack can reach cold temperatures. As a consequence, the stack can shrink for a short time before the burner tip ignites. During this time, a steady-state thermal regime is established. In this case, the cold temperatures would exist in the stack metal in a transient condition. When the flare tip ignites and burns, the temperature of the surroundings and the stack metal elevates. This is

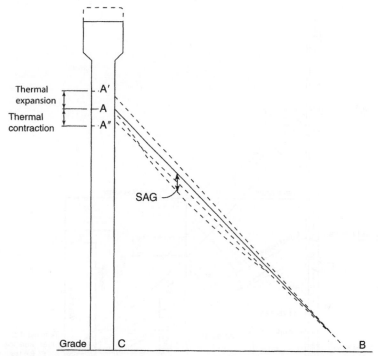

FIGURE 3.10   Flare stack guy wire thermal movements. Points A, A', and A" are at the cable connection on the upper portion of the flare stack; Point C is at the base of the stack; and Point B is where the cable ties into the ground with a dead man. .... The term "dead man" refers to an anchor in the ground that has sufficient rigidity to restrain the forces in the cables. Typically they are concrete masses embedded in the ground.

a fact observed both in the field and by analytical studies. Of course, if a hot fluid enters the stack, the stack will increase in height.

The guy wires therefore must accommodate the various thermal conditions. ASTM SA-333 pipe material (3½ Ni—comparable to BS 3603 HF5503 LT100 CAT.2) is used for the stack. When cold, this material will not experience brittle fracture under wind loads. However, the use of this material does not mean that the stack is always cold when flaring. In many instances, the flare metal temperature is hot and will experience thermal growth. Therefore, the flare must be designed for both cases.

Because the flare stack gets hot and grows in height, the cable has to incorporate sag to accommodate stretching under tension. Figure 3.10 illustrates the general flare scheme with the guy wires.

The guy wire extends from the flare stack ring to the dead man position, forming the hypotenuse of a triangle with the flare stack and the ground. In the nonoperating position, this forms triangle A-B-C. As the flare stack heats up and grows, the distance from the guy cable support ring on the flare stack to

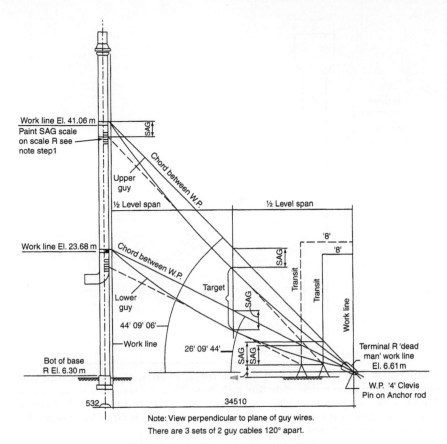

FIGURE 3.11   Flare stack guy cable details.

the dead man increases, making the guy cable stretch. This is shown as triangle A′-B-C. If the flare stack cools and contracts, the guy cable will shrink, shown as A″-B-C. Because the guy wire is forced to stretch (or break) in the hot case (triangle A′-B′C), the cable incorporates sag to compensate for this movement. There is no sag required for the cold case as the guy cable decreases in length, since line A″-C is shorter than A-C. Line A′-C is longer than line A-C and is the worst case for determining the sag. The actual configuration is shown in Figure 3.11.

The analytical expression for the sag of a uniformly loaded beam under tensile loads, shown in Figure 3.12, is given by the following:

$$y_{max} = -\frac{wL^2}{8P} \tag{3.14}$$

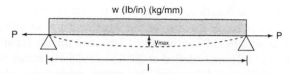

FIGURE 3.12 Beam subjected to a uniform load with axial tensile load.

where
$y =$ in (mm)
$w =$ cable weight (lb_m/in, kg/mm)
$L =$ cable length (in, mm)
$P =$ tensile force (lb_f, N)

The minus sign in Eq. 3.14 indicates that the sag is in the negative direction. As you can see from the equation, the amount of sag is inversely proportional to the tensile force, P. Thus, to theoretically have zero sag would require an infinite amount of force. In practical terms, there will always be some sag. Solving for the tensile force and using a negative value for y_{max} (downward direction), you obtain the following equation:

$$P = \frac{wL^2}{8y_{max}} \qquad (3.15)$$

The amount of sag required depends on the magnitude of thermal expansion of the stack. As stated earlier, if the stack expands upward, the cable is stretched. If the stack contracts, the cable shrinks in length. Thus, the stack expansion upward sets the amount of cable sag.

The equation for the deflection of a cable under its own weight is given by the catenary equation, written as follows:

$$P = \frac{(y_{max})(w)}{\left[\cosh\left(\dfrac{wL}{2P}\right) - 1\right]} \qquad (3.16)$$

It can be shown that for cables (see "Supplement A"), Eq. 3.15 and Eq. 3.16 yield the same results, although each has a different basis of derivation. In the derivation of Eq. 3.16, a similar equation is developed for the arc length of the cable. This is important in establishing the relative influence of the thermal movements of the stack on the cable sag. The arc length of the cable is given as follows:

$$S = \frac{P}{w} \sinh\left(\frac{wL}{2P}\right) \qquad (3.17)$$

Using Eq. 3.17, the length of the cable arc is seen to be almost identical to line A-B in Figure 3.10. In the case of the stack in the field in Saudi Arabia, the distance A-B was 1935.07 inches (49.15 m), and the arc length with a sag of 5.244 inches (133.2 mm; installed at 20°C) is 1935.10 inches (49.152 m): a mere 1/32 inch (0.79 mm) difference! This makes more sense when you consider that the 133.2 mm sag is insignificant compared to 49 m (1935.10 inches). Similarly, if the guy cables are installed at 0°C, the resulting sag is reduced to 4.7 inches (119.38 mm). The resulting tensile load is not greater than the pretension load (see "Supplement B"); therefore, the change in sag is acceptable. With this result, it can be reasonably concluded that the cable sag is approximately equal to the thermal movement of the stack. As the stack heats up, so do the guy cables from thermal radiation induced from the flame exhaust from the flare tip. This is an important assumption because it simplifies the solution process. For this reason, the y_{max} parameter in Eq. 3.15 and Eq. 3.16 was selected to be equal to the sag. Thus, y_{max} can be substituted for the parameter SAG. This is shown in the contractor's drawing in Figure 3.11.

The solution process begins with calculating the thermal growth of the stack at each guy cable ring support. This thermal expansion is then substituted into Eq. 3.15 or Eq. 3.16 to calculate the corresponding tensile force, P. This was carried out with both equations, and the results verified that they yielded the same results. However, Eq. 3.16 is more awkward to use, since solving for P requires a trial-and-error process; whereas Eq. 3.15 is direct substitution. For this reason Eq. 3.15 is the basic equation to be used in this method. However, the use of Eq. 3.16 gives the required mathematical rigor in verifying the engineering mechanics of the problem. Its derivation is simpler than Eq. 3.15 and is published in every elementary textbook on statistics.

To compute the flare stack thermal expansion, you compute the average temperature of the flare stack by taking the average of the temperature at the guy support ring and the temperature at the ground. This is carried out for each elevation of guy cables. To do this, you use the temperature of the air at the time of tensioning to calculate the thermal expansion of the flare stack and, hence, the cable tension. Calculating the length of thermal expansion gives

$$SAG = \Delta l = \alpha L \left(\Delta T \right) \text{in (mm)} \qquad (3.18)$$

where
ΔT = temperature at guy ring—temperature of air at cable tensioning
α = coefficient of thermal expansion at temperature of stack at guy support ring (in/in − °F (mm/mm − °C).

Substituting $\Delta l = y_{max}$ (=SAG) gives the following:

$$P = \frac{wL^2}{8(SAG)} \qquad (3.19)$$

The tensile force, P, is solved for each ambient air temperature to produce cable tensions. The tensile load is increased to allow for any slack in tensioning the cable and, for the cold case, when the stack shrinks. The excessive tensile load allows for the increased sag in the cable if the stack shrinks in the cold state. This excessive tensile force was developed by using a straight line. The tensile force, P, was calculated for the corresponding SAG length at two points: 0°C and 20°C. A straight line was then drawn, of which the loci of points were the values of P. This is an empirical constant, derived from experience. The constant, C, is a linear function, given as follows:

$$C = 1.1429 - 0.001772T \tag{3.20}$$

where $T = $ temperature (°C).

The value for P in Eq. 3.19 is converted to the S.I. (metric SI) as

$$P = \frac{wL^2 (\mathrm{lb_f})}{8(SAG)} \left(\frac{C}{2.2046 \left(\dfrac{\mathrm{lb_f}}{\mathrm{Kg}} \right)} \right) \left(\frac{9.807\,\mathrm{N}}{\mathrm{Kg}} \right) = \frac{0.556 wL^2 C}{SAG}\,\mathrm{N} \tag{3.21}$$

where
$w = \mathrm{lb_f/in}$
$L = \mathrm{in}$
$SAG = \mathrm{in}$
$C = $ is defined in Eq. 3.20
$P = $ Newtons

Example 3.3: Guy Cables

A guy cable is 1935.07 inches long and weighs $0.154\,\mathrm{lb_m/in}$. At operating flare conditions, the guy cable connection on the upper part of the stack (guy cable support ring) is 700°F and at grade the temperature is 500°F. The elevation of the guy cable connection is 114.042 feet. The ambient air temperature is 104°F (40°C). The material of construction of the flare stack is A333 Gr 3.
   Solution:
   The average temperature of the stack from the guy cable support ring connection to grade is calculated as

$$T = \frac{(700 + 500)}{2} = 600°\mathrm{F}\ (316°\mathrm{C})$$

The coefficient of thermal expansion for the stack material at 600°F, with 70°F reference temperature is

$$\alpha = 7.23 \times 10^{-6}\ \frac{\mathrm{in}}{\mathrm{in} - °\mathrm{F}}$$

Solving for the thermal expansion of the stack, you can calculate

$$\Delta l = \alpha l(\Delta T) = (7.23 \times 10^{-6})\frac{in}{in - {}^\circ F}(114.04) in\,(12)\frac{in}{ft}(600 - 70)^\circ F$$

$$\Delta l = 5.244\,in\,(133.2\,mm)$$

Since we are using 70°F (21°C) as the reference temperature, we must compensate for the ambient air at 104°F (40°C). At 104°F the coefficient of thermal expansion of the cable material is 6.1396×10^{-6} in/in$-$°F. Solving for the expansion of the guy cable, you can calculate

$$\Delta l = \alpha l(\Delta T) = (6.14 \times 10^{-6})\frac{in}{in - {}^\circ F}(1935.07) in\,(104 - 70)^\circ F$$

$$\Delta l = 0.404\,in$$

The total sag for the cable considering thermal expansion is

$$SAG = 5.244\,in + 0.404\,in = 5.648\,in = 143.46\,mm$$

Now solving for C in Eq. 3.20, you have

$$C = 1.1429 - 0.001772(20) = 1.1075$$

With total sag of 5.648 inches, the required tension in the guy cable is, from Eq. 3.21, as follows:

$$P = \frac{(0.556)(0.154)(1935.07)^2(1.1075)}{5.648} = 62869.20\,N$$

or $P = 14132.996\,lb_f$

This same process is repeated for ambient temperatures of 0°C, 10°C, 20°C, and 50°C—each representative for different seasons in Saudi Arabia.

Supplement A

The following is a comparison of Eq. 3.15 and Eq. 3.16.
Setting both equations equal to each other

$$\frac{y_{max}w}{\cosh\left(\dfrac{wL}{2P}\right) - 1} = \frac{wL^2}{8y_{max}}$$

This equation results in the following:

$$\cosh\left(\frac{wL}{2P}\right) = 1 + 8\left(\frac{y_{max}}{L}\right)^2$$

Substituting in values for the 20°C case,
W = 0.154 lb$_m$/in, L = 1935.07 in, P = 68943.21, N = 15498.338 lb$_f$, you have

$$\cosh\left(\frac{wL}{2P}\right) = 1.0000462$$

$$1 + 8\left(\frac{y_{max}}{L}\right)^2 = 1.0000588$$

Comparing these results with those of other installation temperatures, you can assume that Eq. 3.15 and Eq. 3.16 yield the same results for this application.

Supplement B

The purpose of this section is to solve for the tensile load and the resulting sag for the case of thermal expansion that results in the cable extending from 1935.07 inches to 1938.557 inches at the installation temperature of 0°C.

Writing Eq. 3.15 and solving for P,

$$P = \frac{wL^2}{8y_{max}}$$

Substituting w = 0.154 lb$_m$/in, L = 1938.557 in, you solve for a variable, Δ, such that

$$\Delta = \frac{wL^2}{8y_{max}} - P = \frac{0.154(1938.557)^2}{8y_{max}} - P$$

or

$$\Delta = \frac{72341.562}{y_{max}} - P$$

For $\Delta = 0$, the theoretical solution is y_{max} = 4.7 inches and P = 15391.8 lb$_f$. Since P is less than the pre-tension load of 76396.53 N = 17,173.834 lb$_f$, you can assume that the elongation of the cable has relatively no effect on the cable sag. The cable sag can be assumed to equal the thermal expansion of the flare stack. This is the basis of that assumption.

Note: After the cables were installed at the location in Saudi Arabia, the flare stack flared for 30 days nonstop.

REFERENCES

1. Escoe, A. Keith, *Mechanical Design of Process Systems,* vol. 2, 2nd edition, Gulf Publishing Company, Houston, Texas, 1995.
2. ASME STS-1-2006, American Society of Mechanical Engineers, New York, NY, 2006.
3. Findlay, Matt, *Improved Screening Criteria for the Prediction of Wind Induced Vibration of Tall Process Vessels Arising from Recent Field Experience*, ASME, New York, NY, November 2001.
4. Thomas, G. B., *Calculus and Analytic Geometry*, 3rd edition, Addison-Wesley Publishing Company, Inc., Reading, Massachusetts, U.S.A., 1960.
5. Escoe, A. Keith, *Guyed Flare Stacks*, Hydrocarbon Engineering, Farnham, Surrey, England, Sept., 1998.

Chapter 4

Wind Loadings on Pressure Vessels and Stacks

This chapter considers wind loads as they relate to field applications. This discussion is not intended to be for the wind design of problems; albeit, the material presented can be used in wind design. This chapter provides a generic guide to wind because, depending on the location of the facility where the analysis is being performed, the wind codes may vary. Some areas do not have formal wind codes as such. Often the wind codes of other countries are applied in these locations—even though the wind isopleths are not applicable. Most often people estimate an appropriate wind speed and use the preferred code accordingly.

Wind and seismic codes change regularly, and the trend is a shift toward the use of the Load and Resistance Factor Design (LRFD) approach versus the Allowable Stress Design (ASD) basis. The Canadian codes have done this, making their application to stacks and pressure vessels virtually impractical. Some of the bodies that produce the wind standards argue that their documents were never intended for pressure vessels and stacks; their main concern is buildings and civil structures like bridges. This may lead to the development of a separate wind and seismic code for pressure vessels and stacks. However, for the present, we must use what is available.

The consideration of wind loads in a plant or operating facility normally happens when a section of a tower is to be stress-relieved by post weld heat treatment and an entire shell portion of the vessel is to be heated to the stress-relieving temperature. In this application, you must be cognizant of the imposing wind loads on the vessel. Another application would be the installation of guy cables, as discussed in Chapter 3, and the loads on these cables. These situations are routine in plants. In the case of relieving stress on a process column, only the wind shear and bending moments at the location under consideration are of interest.

ASME STS-1 [1] follows the latest ASCE 7 standard series—presently ASCE 7-2005 [2]. We will develop a general guideline of this standard that can be used throughout the world in various locations that do not have wind codes. If you are located in a country that has such a standard, it is advised that you follow that document. However, for operational applications, the methodology proposed here should be adequate.

The general equation for the wind force imposed on a pressure vessel or columns is

$$F = q_z G C_f A_f \, (\text{lb}_f \text{ or N}) \tag{4.1}$$

where
q_z = velocity pressure at elevation of height z of the centroid of area A_f
G = gust-effect factor
C_f = force coefficient
A_f = projected area normal to the wind

In this chapter we will discuss the terms in Eq. 4.1 and guidelines on their application.

THE VELOCITY PRESSURE DISTRIBUTION, q_z

The term q_z is the velocity pressure and is evaluated at a height z. It is defined as follows:

$$q_z = 0.00256 K_z K_{zt} K_d V^2 I \, (\text{lb}_f/\text{ft}^2) \tag{4.2}$$

or

$$q_z = 0.613 K_z K_{zt} K_d V^2 I \quad (\text{N/m}^2); \quad V \text{ is in m/s} \tag{4.2a}$$

where
K_d = wind directionality factor
K_z = velocity pressure exposure coefficient
K_{zt} = topographic factor

WIND DIRECTIONALITY FACTOR, K_d

The wind directionality factor, K_d, is defined in Table 4.1 (Table 6.4 of ASCE 7-2005 [2]).

TABLE 4.1 Wind Directionality Factor, K_d

Structure type—chimneys, tanks, and similar structures	Directionality factor, K_d
Square	0.90
Hexagonal	0.95
Round	0.95

(Courtesy of the American Society of Civil Engineers)

VELOCITY PRESSURE COEFFICIENT, K_z

The velocity pressure coefficient factor, K_z, is determined from Table 4.2 (Table 6.3 of ASCE 7-2005 [2]).

TABLE 4.2 K_z and Exposure Coefficients

Height above ground level, z		Exposure category		
ft	m	B	C	D
0–15	0–4.6	0.57	0.85	1.03
20	6.1	0.62	0.9	1.08
25	7.6	0.66	0.94	1.12
30	9.1	0.7	0.98	1.16
40	12.2	0.76	1.04	1.22
50	15.2	0.81	1.09	1.27
60	18	0.85	1.13	1.31
70	21.3	0.89	1.17	1.34
80	24.4	0.93	1.21	1.38
90	27.4	0.96	1.24	1.4
100	30.5	0.99	1.26	1.43
120	36.6	1.04	1.31	1.48
140	42.7	1.09	1.36	1.52
160	48.8	1.13	1.39	1.55
180	54.9	1.17	1.43	1.58
200	61	1.2	1.46	1.61
250	76.2	1.28	1.53	1.68
300	91.4	1.35	1.59	1.73
350	106.7	1.41	1.64	1.78
400	121.9	1.47	1.69	1.82
450	137.2	1.52	1.73	1.86
500	152.4	1.56	1.77	1.89

(Courtesy of the American Society of Civil Engineers)

TABLE 4.3 Terrain Exposure Constants

Exposure	α	z_g (ft)	Z_{min} (ft)
B	7.0	1200	30
C	9.5	900	15
D	11.5	700	7

(Courtesy of the American Society of Civil Engineers)

TABLE 4.4 Terrain Exposure Constants (Metric)

Exposure	α	z_g (m)	Z_{min} (m)
B	7.0	365.76	9.14
C	9.5	274.32	4.57
D	11.5	213.36	2.13

(Courtesy of the American Society of Civil Engineers)

The velocity pressure coefficient, K_z, may be determined from the following:

For $15\,\text{ft} \leq z \leq z_g$

$$K_z = 2.01 \left(\frac{z}{z_g} \right)^{2/\alpha} \tag{4.3}$$

For $z > 15\,\text{ft}$

$$K_z = 2.01 \left(\frac{15}{z_g} \right)^{2/\alpha} \tag{4.4}$$

The constants α and z_g are tabulated in the U.S. Customary system in Table 4.3 and in the metric SI in Table 4.4.

TOPOGRAPHIC FACTOR, K_{zt}

The topographic factor, K_{zt}, is perhaps one of the most difficult to define, especially in remote locations. In remote locations in many regions of the world, it would be undesirable to stake out the terrain for many reasons—safety, accessibility, ability to record changes in terrain. Paragraph 6.5.7.2 of ASCE 7-2005

[2] states, "If site conditions and locations of structures do not meet all the conditions specified in Section 6.5.7.1 then $K_{zt} = 1.0$." The wind exposure categories are defined in Paragraph 6.5.6.2 [2] as a function of the surface roughness, as follows:

Surface Roughness B: Urban and suburban areas, wooded areas or other terrain with numerous closely spaced obstructions having the size of single-family dwellings or larger.

Surface Roughness C: Open terrain with scattered obstructions having heights generally less than 30 ft (9.1 m). This category includes flat open country, grasslands, and all water surfaces in hurricane-prone regions.

Surface Roughness D: Flat, unobstructed areas and water surfaces outside hurricane-prone regions. This category includes smooth mud flats, salt flats, and unbroken ice.

If the topographic factor defined in Paragraph 6.5.7 [2] cannot adequately be determined, and if there is no specification, such as from a client, normally it is adequate to be conservative on the Surface Roughness Category and use $K_{zt} = 1.0$ in the field.

BASIC WIND SPEED, V

The basic wind speed, V, is often found in the isopleth charts of the given wind code for a certain area. Where there is no such standard for the area concerned, you can use various readings and communication with the locals to find a conservative wind speed value. In desert regions, sand storms, known as shamals, can develop and appear suddenly if there are no mass communication warnings from a weather station. Normally, a shamal can be seen approaching from a long distance, so there is some warning. Some shamals can reach hurricane-force winds, so you must use judgment if an area has a history and occurrence of these storms in desert regions.

IMPORTANCE FACTOR, I

For process columns and stacks, the importance factor, I, is always made 1.0 because of the importance of the tower.

We have now discussed the terms for solving Eq. 4.2. Now we turn our attention to solving Eq. 4.1.

GUST-EFFECT FACTOR, G

The gust-effect factor, G, is perhaps the most demanding parameter to calculate, and the results are nearly always in the same predictable range. Paragraph 6.5.8.1 of ASCE 7-2005 [2] defines a *rigid structure* as one in which the structure's fundamental natural frequency is equal to or greater than 1 Hz. For a rigid structure, $G = 0.85$. A *flexible structure* is one in which the fundamental natural frequency is less than 1 Hz. In this case, you must perform a very long algorithm to obtain G. Normally, G is seldom less than 1.4. If you are making a quick estimate and want to be conservative, a value of $G = 1.5$ will handle the vast majority of cases. This is a rule of thumb that can be helpful.

Solving for the Gust-Effect Factor, *G*

The equation for G is given in Paragraph 6.5.8.2 of ASCE 7-2005 [2] for flexible or dynamically sensitive structures. This equation, listed as Eq. 6.8 in the standard, reads as follows:

$$G_f = 0.925 \left(\frac{1 + 1.7 I_{\bar{z}} \sqrt{g_Q^2 Q^2 + g_R^2 R^2}}{1 + 1.7 g_v I_{\bar{z}}} \right) \tag{4.5}$$

where, in Eq. 4.1, $G = G_f$ for a flexible structure.

The background response, Q, is given by the following:

$$Q = \sqrt{\frac{1}{1 + 0.63 \left(\frac{OD + h}{L_{\bar{z}}} \right)^{0.63}}} \tag{4.6}$$

where
OD = vessel or stack outside diameter (ft) (denoted as B in ASCE 7-2005 [2])
h = height of vessel or stack (ft)

$$L_{\bar{z}} = l \left(\frac{\bar{z}}{33} \right)^{\bar{\varepsilon}} \tag{4.7}$$

where l and $\bar{\varepsilon}$ are given in the U.S. Customary system in Table 4.5 and in the metric SI in Table 4.6 (taken from the ASME STS-1 2006 Table I.1 [1]).

Setting f_1 = fundamental natural frequency of the tower (ASCE 7-2005 [2] uses n_1), V = basic wind speed, mph (m/s), in which $80 \le V \le 140$ mph (some note as V_{ref}).

ASCE 7-2005 [2] lists the following formulations to solve Eq. 4.5:

$$\overline{V_{\bar{z}}} = \bar{b} \left(\frac{\bar{z}}{33} \right)^{\bar{\alpha}} V \left(\frac{22}{15} \right) \tag{4.8}$$

TABLE 4.5 Gust-Effect Parameters

Exposure	l (ft)	$\bar{\varepsilon}$	c	$\bar{\alpha}$	$\bar{b}$
B	320	30	0.30	1/4.0	0.45
C	500	15	0.20	1/6.5	0.65
D	650	7	0.15	1/9.0	0.80

(Courtesy of the American Society of Civil Engineers)

TABLE 4.6 Gust-Effect Parameters (Metric)

Exposure	l (m)	$\bar{\varepsilon}$	c	$\bar{\alpha}$	$\bar{b}$
B	97.54	9.14	0.30	1/4.0	0.45
C	152.4	4.57	0.20	1/6.5	0.65
D	198.12	2.13	0.15	1/9.0	0.80

(Courtesy of the American Society of Mechanical Engineers)

$$R_l = R_h \quad \text{where } R_h = \frac{1}{\eta_1} - \frac{1}{2\eta_1^2}(1 - e^{-2\eta_1}) \tag{4.9}$$

$$R_l = R_B : R_B = \frac{1}{\eta_2} - \frac{1}{2\eta_2^2}(1 - e^{-2\eta_2}), \quad \text{where } B = OD \tag{4.10}$$

$$R_l = R_d \quad \text{where } R_d = \frac{1}{\eta_3} - \frac{1}{2\eta_3^2}(1 - e^{-2\eta_3}) \tag{4.11}$$

where

$$\eta_1 = \frac{4.6 f_1 h}{\bar{V}_{\bar{z}}} \tag{4.12}$$

$$\eta_2 = \frac{4.6 f_1 (OD)}{\bar{V}_{\bar{z}}} \tag{4.13}$$

$$\eta_3 = \frac{4.6 f_1 d}{\bar{V}_{\bar{z}}} \tag{4.14}$$

where d = top one third of tower (see ASME STS-1 2006, p. 87 [1])

$$N_1 = \frac{f_1 L_{\bar{z}}}{\bar{V}_{\bar{z}}}$$ (4.15)

$$R_n = \frac{7.47 N_1}{(1 + 10.3 N_1)^{5/3}}$$ (4.16)

$$R = \sqrt{\left(\frac{1}{\beta}\right) R_n R_h R_B (0.53 + 0.47 R_d)}$$ (4.17)

where β = damping ratio $\rightarrow 0.002 \le \beta \le 0.004$
  The intensity of turbulence at height z is $I_{\bar{z}}$, where

$$I_{\bar{z}} = c\left(\frac{33}{z}\right)^{1/6}$$ (4.18)

where $\bar{z} = 0.6\,h$ (4.19)

                          = the equivalent height of the tower

$\bar{z} \ge z_{min}$, where z_{min} is listed in Tables 4.3 and 4.4 and c in Eq. 4.18 is listed in Tables 4.5 and 4.6
  Now

$$g = 32.2\,\frac{ft}{sec^2}; \quad g_Q = g_v = 3.4$$

$$g_R = \sqrt{2\,Ln(3600 f_1)} + \frac{0.577}{\sqrt{2\,Ln(3600 f_1)}}$$ (4.20)

    Now you have the equations to solve Eq. 4.5. Paragraph 6.5.8.3 of ASCE 7-2005 [2] notes, "In lieu of the procedure defined in 6.5.8.1 and 6.5.8.2 [the above algorithms for solving for the gust-effect parameter], determination of the gust-effect factor by any rational analysis defined in the recognized literature is permitted." We have already given rules of thumb for this parameter based on proven engineering practice.

THE PROJECTED AREA NORMAL TO THE WIND, A_f

The final parameter to solve for in Eq. 4.1 is the projected area normal to the wind, A_f. Figure 4.1 shows the effective wind diameter for a process column.

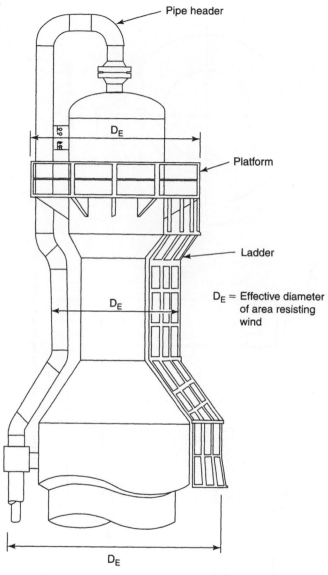

FIGURE 4.1   The effective wind diameter can vary with height [3].

As shown in the figure, the effective wind diameter can vary with height. Figure 4.2 shows the effective wind diameter of a conical section.

To calculate the parameter A_f, you multiply D_E by the length of the section. Now you have the tools to solve Eq. 4.1.

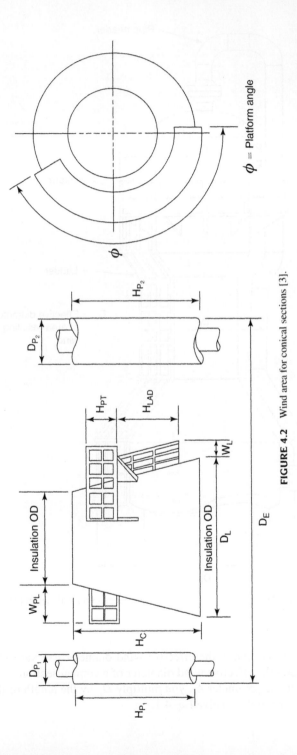

FIGURE 4.2 Wind area for conical sections [3].

EXAMPLE 4.1

In this example, a tower that has the same geometry as that in Example 3.1 is made of a different metallurgy and has a different weight distribution. Because of the metallurgy of the tower, any welding requires post weld heat treating (PWHT) of the section having hot work performed for an upcoming turnaround. When performing PWHT, you need to know the wind loads exerted on the subject tower section. The areas of repair will be determined by the turnaround team.

The facility is located on the Texas gulf coast, and the maximum expected basic wind speed during the PWHT is 85 mph, although the wind zone is 120 mph (for this example, do not expect to perform such an operation during a hurricane, so 85 mph is the maximum expected basic wind speed for analysis; the actual basic wind speed was 30 mph). This application is for an operations facility, not an engineering firm that is designing a tower for the worst possible conditions.

What you need is a chart of wind pressures and shear and bending moments for the turnaround.

The ASCE 7-2005 [2] parameters are as follows:

Wind force coefficient, $C_f = 0.7$
Basic wind speed, $V = 85$ mph
Importance factor, $I = 1$
Exposure category $= C$
Wind Directionality Factor, $K_d = 0.95$
Topographic Factor, $K_{zt} = 1.00$

Vessel Characteristics:
Vessel height, $h = 80.4152$ ft
Vessel minimum diameter, $b = OD = 2.0417$ ft
Corrosion allowance $= 0$
Fundamental Frequency $= n_1 = f_1 = 0.9534$ Hz
Damping coefficient for operating condition, $\beta = 0.0191$

Basic Load Combinations for Allowable Stress Design (ASD):
$D + H + W$
$0.6D + H + W = 0.6D + H + W$

where

$D =$ dead load
$H =$ Pressure load
$W =$ Wind load

Gust-Factor Calculations for the Operating Condition:

$$\bar{z} = 0.6h$$
$$= 0.60(80.4152)$$
$$= 48.2491$$

$$I_{\bar{z}} = c\left(\frac{33}{\bar{z}}\right)^{1/6}$$

$$= 0.200\left(\frac{33}{48.2491}\right)^{1/6}$$

$$= 0.1877$$

$$L_{\bar{z}} = l\left(\frac{\bar{z}}{33}\right)^{\bar{\varepsilon}} = 500\left(\frac{48.2491}{33}\right)^{0.200}$$

$$= 539.4673$$

$$Q = \sqrt{\frac{1}{\left[1 + 0.63\left(\dfrac{h + OD}{L_{\bar{z}}}\right)^{0.63}\right]}}$$

$$Q = 0.9156$$

$$V_{\bar{z}} = \bar{b}\left(\frac{\bar{z}}{33}\right)^{\bar{\alpha}} V_{ref}\left(\frac{88}{60}\right)$$

$$= 0.6500\left(\frac{48.2491}{33}\right)^{0.1538}(85)\left(\frac{88}{60}\right)$$

$$= 85.9102$$

$$N_1 = f_1\left(\frac{L_{\bar{z}}}{V_{\bar{z}}}\right) = 0.9534\left(\frac{539.4673}{85.9102}\right) = 5.9867$$

$$R_n = \frac{7.465 N_1}{\left(1 + 10.302 N_1\right)^{5/3}} = \frac{7.465\,(5.9867)}{\left(1 + 10.302\,(5.9867)\right)^{5/3}} = 0.0452$$

$$\eta_1 = \frac{4.60 f_1 h}{V_{\bar{z}}} = \frac{4.60\,(0.9534)\,(80.4152)}{85.9102} = 4.1050$$

$$R_h = \frac{1}{n_1} - \frac{1 - e^{-2n_1}}{2n_1^2} = \frac{1}{4.1050} - \frac{1 - e^{-2(4.1050)}}{2(4.1050)^2} = 0.2139$$

$$\eta_2 = \frac{4.60 f_1 \, (OD)}{V_{\bar{z}}} = \frac{4.60(0.9534)(2.0417)}{85.9102} = 0.1042$$

$$R_B = \frac{1}{\eta_2} - \frac{1 - e^{-2\eta_2}}{2\eta_2^2} = \frac{1}{0.1042} - \frac{1 - e^{-2(0.1042)}}{2(0.1042)^2} = 0.9340$$

$$\eta_3 = \frac{15.40 f_1 d}{V_{\bar{z}}} = \frac{15.40(0.9543)(2.0417)}{85.9102} = 0.3489$$

$$R_d = \frac{1}{\eta_3} - \frac{1 - e^{-2\eta_3}}{2\eta_3^2} = \frac{1}{0.3489} - \frac{1 - e^{-2(0.3489)}}{2(0.3489)^2} = 0.8029$$

$$R = \sqrt{\left(\frac{1}{\beta}\right) R_n R_h R_B (0.53 + 0.47 R_d)}$$

$$R = \sqrt{\left(\frac{1}{0.0191}\right)(0.0452)(0.2139)(0.9340)(0.53 + 0.47(0.8029))} = 0.4286$$

$$g_R = \sqrt{2 \mathrm{Ln}(3600 f_1)} + \frac{0.577}{\sqrt{2 \mathrm{Ln}(3600 f_1)}}$$

$$g_R = \sqrt{2 \mathrm{Ln}(3600(0.9534))} + \frac{0.577}{\sqrt{2 \mathrm{Ln}(3600(0.9534))}} = 4.1781$$

$$G_f = \frac{0.925(1 + 1.7 I_{\bar{z}} \sqrt{g_Q^2 Q^2 + g_R^2 R^2})}{(1 + 1.7 g_v I_{\bar{z}})}$$

$$G_f = \frac{0.925\left(1 + 1.7(0.1877)\sqrt{(3.40)^2 0.8383 + (4.1781)^2 (0.4286)}\right)}{(1 + 1.7(3.40)(0.1877))} = 1.0303$$

Wind Pressure Calculations:

$$K_z = 2.01 \left(\frac{z}{z_g}\right)^{2/\alpha} = 2.01 \left(\frac{z}{900}\right)^{0.2105}$$

$$q_z = 0.00256 K_z K_{zt} K_d V^2 I = 0.00256 K_z (1.0)(0.950)(85.0)^2 (1.0)$$

$$WP = q_z G_f C_f \left(\text{Minimum of } 10 \frac{lb_f}{ft^2}\right)$$

$$WP = q_z (1.0303)(0.7) = 0.7212 q_z$$

Table 4.7 lists the wind pressures for the job site
Table 4.8 shows the wind loadings that the turnaround team requested. From this table, you can quickly determine the stresses imposed by the wind when the proper areas of repair and PWHT are determined.

TABLE 4.7 Wind Pressures

Height Z (ft)	K_z	q_z (psf)	WP: Operating (psf)
15.0	0.8489	14.92	10.76
20.0	0.9019	15.85	11.43
25.0	0.9453	16.61	11.98
30.0	0.9823	17.26	12.45
40.0	1.0436	18.34	13.22
50.0	1.0938	19.22	13.86
60.0	1.1366	19.97	14.40
70.0	1.1741	20.63	14.88
80.0	1.2075	21.22	15.30
90.0	1.2379	21.75	15.69

TABLE 4.8 Wind Loadings and Deflection Report

Component	Elevation from above Base (in)	Effective OD (ft)	Elastic modulus E(10⁶) psi	Inertia (ft⁴)	Total wind Shear at Bottom (lbf)	Bending Moment at Bottom (ft-lbf)	Deflection at top (in)
Ellipsoidal Head	953.67	3.07	29.0	*	36.54	14.97	2.2417
Cylinder #1	791.67	3.05	29.0	0.2834	661.82	4753.67	2.1984
Transition #1	779.67	2.57	29.0	*	700.06	5435.85	1.5800
Cylinder #2	575.67	2.04	29.0	0.0675	1202.15	21651.48	1.5344
Cylinder #3	455.67	2.06	29.0	0.1029	1485.19	35099.30	0.8221
Cylinder #4	359.67	2.08	29.0	0.1393	1705.39	47862.49	0.4954
Cylinder #5	263.67	2.10	29.0	0.1769	1911.91	62339.17	0.2937
Transition #2	146.05	2.39	29.0	*	2159.39	82186.05	0.1480
Cylinder #6	39.70	2.67	29.0	0.5648	2388.89	102242.47	0.0465
Ellipsoidal Head #4 (top)	35.00	2.64	29.0	*	2399.90	103180.23	0.0044
Support Skirt	0.00	2.61	29.3	0.3765	2481.94	110299.57	0.0035

REFERENCES

1. *ASME STS-1-2006*, American Society of Mechanical Engineers, 2006, New York, NY.
2. ASCE, American Society of Civil Engineers, ASCE 7-2005, *Minimum Design Loads for Buildings and Other Structures*, New York, NY, 2005.
3. Escoe, A. Keith, *Mechanical Design of Process Systems,* vol. 1, 2nd edition. Gulf Publishing Company, Houston, Texas, 1995.

Chapter 5

Pressure Vessel Internal Assessment

The internals of pressure vessels are what make the vessels work. The processing of fluids is the integral function of these pressure vessels. As a result, the failure of internals affects the performance of any operating facility. The consequence of such failures normally results in unscheduled shutdowns, depending on the severity of the damage. Not much has been written on this subject because people do not want to admit to such failures—let alone advertise the event. Thus, this chapter was difficult to write but is based on actual cases. This chapter does not address the design of internals, but rather the assessment of the mechanisms that cause failures of internals. The most common failure of the internals of pressure vessels is fatigue. Figure 5.1 shows a fatigue failure that initiated the collapse of the internals of an MTBR Regenerator air distributor.

The failure shown in Figure 5.1 involved more than just fatigue alone. In a large fluidized bed, pulsation of the internal fluids initiates strong and rapidly changing forces. For such services, the internals must be designed adequately

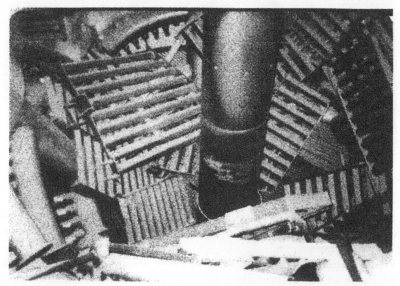

FIGURE 5.1   The collapse of internals of an MTBR Regenerator air distributor.

FIGURE 5.2   Air distributor piping.

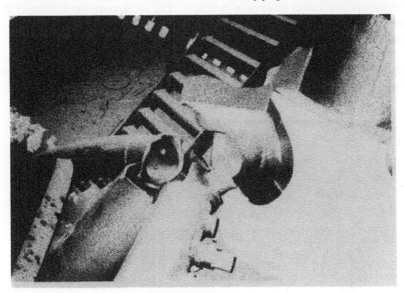

FIGURE 5.3   Gusset plate attachment.

for these conditions. In this failure, the large pulsation responses throughout the fluidized bed ripped out trays and other internals, and they were piled up at the bottom of the vessel.

Figure 5.2 shows air distributor piping in this scenario. Figure 5.3 provides a closer view of the fatigue failure. Here, a fatigue crack initiated in a gusset plate attachment set up the fatigue crack that propagated and allowed the air distributor pipe to break completely.

FIGURE 5.4 Bent and warped channel sections.

The fatigue crack in the gusset plate attachment resulted in large stress concentrations. With a highly cyclic service, the fatigue crack initiated and resulted in the eventual failure of the air distributor pipe. The use of abrupt geometric boundaries on surfaces is not good practice in cyclic service because of stress intensification.

The effect of the large pulsation forces is graphically shown in Figure 5.4. The bent channels and warped sections resulted from the powerful pressure pulsations extending up and down the regenerator.

The only member of this vessel left standing upright is the center pipe shown on the far right in Figure 5.4. Everything attached to this center pipe was ripped loose—including the thermowells on the walls.

Figure 5.5 shows warped reinforcing sections, a result of the forces described in the preceding paragraphs.

Regarding the figures shown here, there was argument between the parent company and consultants as to what initiated such failures. The parent company thought that thermal gradients were the root cause; however, lab tests confirmed the failure shown in Figure 5.3 was due to fatigue.

During the rebuild, forged fittings with smooth surface contours and self-reinforced nozzle designed for a cyclic surface, as shown in Figure 5.6, replaced gusset plates shown in Figures 5.2 and 5.3.

FORCES ON INTERNAL COMPONENTS

Forces exerted on internal components can be controlled with proper operational procedures. However, if procedures are violated and/or mistakes are made, such internal forces can be multiplied in magnitude.

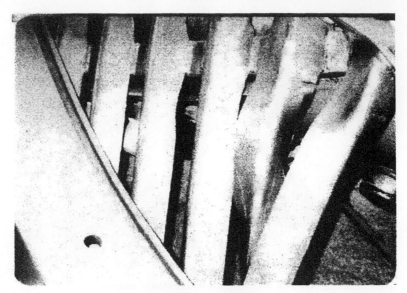

FIGURE 5.5 Warped reinforcing members.

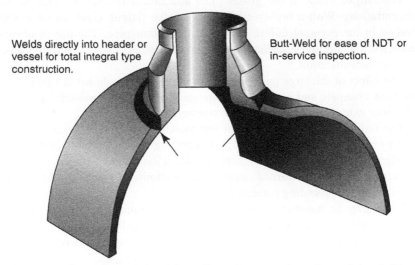

Welds directly into header or vessel for total integral type construction.

Butt-Weld for ease of NDT or in-service inspection.

FIGURE 5.6 Forged fitting designed for cyclic service was an integral part of the rebuild and withstood the pulsation pressure forces (courtesy of WFI International, Inc.).

As you can see in the previous figures, fluid forces acting on internals can be enormous. Typically, column internals are designed to accommodate the weight of at least one man, depending on the diameter of the column. A typical value would be a downward force of 1000 Newtons. The weight of fluid on trays in a column would be equivalent to twice the weir height with

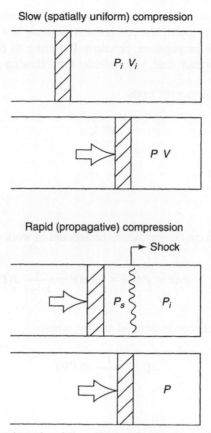

Slow (spatially uniform) compression

$P_i V_i$

$P V$

Rapid (propagative) compression

→ Shock

P_s P_i

P

FIGURE 5.7 Comparison of spatially uniform compression to propagative compression.

a stipulated minimum hydrostatic weight per unit area of the trays. A typical minimum value would be 1 KPa (0.001 MPa) or 0.145 psi. The upward and downward pressures induced by the process conditions must be stipulated by the process engineers or licensor. Typical values for upward and downward pressures would be approximately 2 KPa.

The deflection of sieve, valve, grid, and bubble trap trays is held to a minimum to minimize disruption of the process. One typical value is 1/800 times the column diameter with a maximum of 6 mm.

In fluidized bed reactors and regenerators, the magnitude of compressed gas pressure varies whether the compression is made slowly or rapidly. A gas that is compressed slowly such that the pressure rises uniformly in a control volume is known as *slow*, or *spatially uniform*, *compression*. A rapidly compressed gas, such as by a rapid piston motion, is known as *propagative compression*. Figure 5.7 shows the schematics of both spatially uniform and propagative compression.

Moody [1] has shown that, for a gas to be compressed slowly so that its pressure rises uniformly in a cylinder, the compressed pressure is 43% of the pressure obtained by rapid, or propagative, pressure. Referring to Figure 5.7, the final pressure of the gas for each case, you consider the following parameters:

$k = 1.4 =$ gas ratio of specific heats

$$\frac{V}{V_i} = \frac{1}{2}$$

$Pi = 1.0$ atmosphere
$Ps = 10$ atmospheres
$F = PA$

For spatially uniform compression, the amount of work done is

$$dW_k = Fdx = PAdx = -PdV = \frac{1}{k-1}d(PV)$$

The last term is the change in internal energy, where

$$dU = \frac{1}{k-1}d(PV)$$

Now,

$$\int_{P_i}^{P}\frac{dP}{P} = -k\int_{V_i}^{V}\frac{dV}{V}$$

$$\frac{P}{P_i} = \left(\frac{V_i}{V}\right)^k = (2)^{1.4} = 2.6$$

Likewise, for rapid, or propagative, compression, you have the following:
$Fs = PsA$

$$dW_k = F_s dx = P_s Adx = P_s dV = \frac{1}{k-1}d(PV)$$

where

$$dU = \frac{1}{k-1}d(PV)$$

Now,

$$\int_{P_i}^{P} \frac{dP}{P + (k+1)P_s} = -\int_{V_i}^{V} \frac{dV}{V}$$

$$\frac{P}{P_i} = \frac{V}{V_i} + (k-1)\frac{P_s}{P_i}\left(\frac{V_i}{V} - 1\right)$$

$$\frac{P}{P_i} = 2 + (1.4 - 1)(10)(2 - 1) = 6$$

Thus, you can see that the propagative compression results in a final pressure are 2.3 times that produced by spatially uniform compression. There is the added effect of shock with propagative compression that can enhance forces exerted on internal vessel components.

LINED PLATES AND INTERNAL COMPONENTS

Vessels are lined with another material to provide corrosion resistance. The lining we are addressing here is metal clad plates. These plates have the following features:

1. Weld overlaid.
2. They are integrally clad by explosion welding.
3. They are produced from a roll-bonded integrally clad plate by forming into a cylinder and welding.

In a vessel's operating life, there will be shutdowns and startups. During each of these cycles, the base metal and clad will experience different thermal movements, resulting in shear stresses. These stresses vary as to the relative coefficient of thermal expansion of the two metals. This phenomenon can be analyzed through fatigue analysis, which we will not go into here. The only justification for such an analysis is if many shutdown and startup cycles are expected, resulting in cyclic service. The vast majority of time this phenomenon is not a concern in the operating unit.

Most licensors of various processes have limitations on loadings on clad metal surfaces. A typical requirement is to limit a structural support welded to a clad surface to 13 mm in thickness, with a maximum stress induced by the loading to 34.48 MPa (5 ksi).

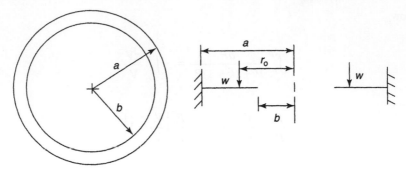

FIGURE 5.8 Tray support ring.

HELPFUL STRUCTURAL FORMULATIONS

The following are formulations for various structural attachments.

Tray Support Ring

Figure 5.8 illustrates a tray support ring. The required thickness and resulting deflection of the support ring are derived from Roark [2]. These equations are based on the material having a Poisson ratio of 1/3. For more comprehensive equations, including a different Poisson ratio, refer to Roark [2]. The tray support ring thickness equation and the maximum deflection equations at the ring center are as follows:

$$t_r = \sqrt{\frac{3 r_o w}{2 S_r}\left(1 - \frac{r_o^2}{a^2}\right) + \frac{18 M}{S_r}\left(\frac{b^2}{a^2 + 2b^2}\right)} \qquad (5.1)$$

where,
t_r = required support ring thickness, not including corrosion allowance, in, mm
a, b, r_o are defined in Figure 5.8
S_r = allowable stress at temperature, psi (MPa)
w = Force per unit length at a point on the ring. Shown on the right side of Figure 5.8 is the axisymmetric representation of the support ring, lb_f/in (N/mm).

$$M = \frac{w r_o}{6}\left[2 LN\left(\frac{a}{r_o}\right) + \frac{r_o^2}{a^2} - 1\right] \qquad (5.2)$$

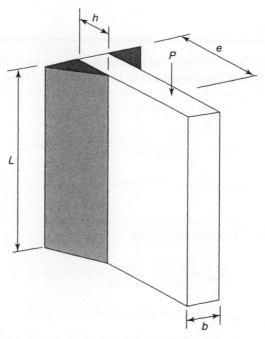

FIGURE 5.9   Support clip welded on two sides with fillet welds; see Giachino et al. [3].

Support Clip Welded on Two Sides with Fillet Welds with Force on Short End

A welded clip is illustrated in Figure 5.9. The weld stresses are as follows:

$$\sigma_{\max} = \frac{4.24Pe}{hL^2}, \text{psi (MPa)} \tag{5.3}$$

$$\tau_{\text{avg}} = \frac{0.707P}{hL}, \text{psi (MPa)—Average shear stress in weld} \tag{5.4}$$

From the von Mises theory, the combined stress is

$$\sigma_c = \sqrt{\sigma_{\max}^2 + 3\tau^2} \tag{5.5}$$

Hicks [4] recommends the following for computing weld stresses:

$$\sigma_c = \beta\sqrt{\sigma_{\max}^2 + 3\tau^2} \tag{5.6}$$

where $0.8 \le \beta \le 0.9$.

Example 5.1

The following is an example of a design of a support clip with fillet welds with a force applied at the short end of the clip.
A clip with L = 127 mm, h = 6 mm, P = 31,000 N, e = 76 mm.
  The maximum stress in the welds is from Eq. 5.3:

$$\sigma_{max} = \frac{4.24(31,000)N(76)mm}{(6)mm\,(127)^2\;mm^2} = 103.2\,MPa$$

$$\sigma_{allow} = 0.6\sigma_y = 0.6(248)MPa = 148.8\,MPa$$

where σ_y = specified minimum yield strength, psi (MPa)
  Now, from Eq. 5.4, you can calculate

$$\tau_{avg} = \frac{0.707(31,000)N}{(6)mm\,(127)mm} = 28.8\,MPa$$

$$\tau_{allow} = (0.4)\sigma_y = (0.4)(248)MPa = 99.2\,MPa$$

$$\sigma_c = 0.9\sqrt{103.2^2 + 3(28.8)^2} = 103.2\,MPa$$

$$\sigma_c^{allow} = 0.6\sigma_y = 0.6(248)MPa = 148.8\,MPa$$

With σ_{max}, τ_{avg}, σ_c are below their respective allowable stresses, and the 6 mm fillet welds an each side of the plate are satisfactory. The plate should be 2 mm (1/16") greater than the weld size, so an 8 mm thick plate should suffice. With a corrosion rate of 3 mm, the actual clip plate thickness is 11 mm.
  Note: The total load is transmitted through the fillet welds only, and no credit is given for possible bearing between the lower part of the clip and the vessel wall. The assumption that the shear, t_{avg}, is carried uniformly in the welds is ubiquitously accepted.
  You can find the derivation of these equations in Bednar, pp. 268–269 [5].

Support Clip with Applied Tensile Force

A welded clip is shown in Figure 5.10.

$$\sigma = \frac{P}{\left(\dfrac{h}{\cos 45°}\right)L} = \frac{0.707P}{hL}\;psi\;(MPa) \tag{5.7}$$

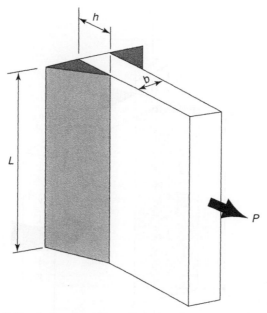

FIGURE 5.10   Support clip with applied tensile force; see Giachino et al. [3].

where

the weld throat size, $t = \dfrac{h}{\cos 45°}$ in (mm) for each fillet weld

h = leg size of the fillet weld, in (mm)

See the note in the preceding section. Also, for the derivation of these equations, see Bednar, pp. 268–269 [5].

Support Clip with Out-of-Plane Bending Moment

A support clip with an out-of-plane bending moment is shown in Figure 5.11.

$$\sigma = \frac{1.414M_{yy}}{hL(b + h)} \text{ psi (MPa)} \tag{5.8}$$

Support Clip with an In-Plane Bending Moment

A support clip with an in-plane bending moment is shown in Figure 5.12.

$$\sigma = \frac{4.24M_{zz}}{hL^2} \text{ psi (MPa)} \tag{5.9}$$

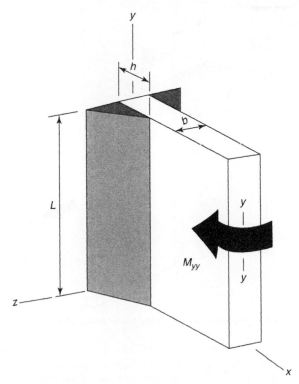

FIGURE 5.11 Support clip with an out-of-plane bending moment, M_{yy}; see Giachino et al. [3].

Support Clip with Continuous Fillet Weld with an Out-of-Plane Bending Moment, M_{yy}

A support clip with continuous fillet weld with an out-of-plane bending moment is shown in Figure 5.13.

$$\sigma = \frac{4.24 M_{yy}}{h\left[b^2 + 3L(b + h)\right]} \text{ psi (MPa)} \tag{5.10}$$

Support Clip with Continuous Fillet Weld with an In-Plane Bending Moment, M_{zz}

A support clip with continuous fillet weld with an in-plane bending moment is shown in Figure 5.14.

$$\sigma = \frac{4.24 M_{zz}}{h\left[L^2 + 3b(L + h)\right]} \text{ psi (MPa)} \tag{5.11}$$

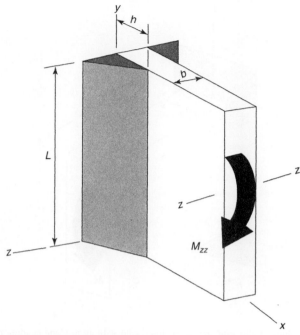

FIGURE 5.12   Support clip with an in-plane bending moment, M_{zz}; see Giachino et al. [3].

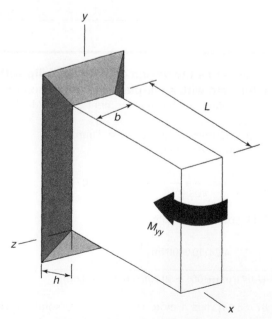

FIGURE 5.13   Support clip with continuous fillet weld with an out-of-plane bending moment, M_{yy}; see Giachino et al. [3].

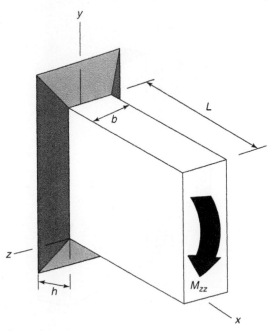

FIGURE 5.14 Support clip with continuous fillet weld with an in-plane bending moment, M_{zz}; see Giachino et al. [3].

Example 5.2

An example illustrating the design of a support clip with continuous fillet weld with an in-plane bending moment.
A support clip has an in-plane bending moment of 50 kNM. The clip has the following parameters:

L = 300 mm; b = 120 mm; weld size = t = 5 mm

Now, M = 50 KNm = 50,000 Nm

$$h = \frac{t}{\cos 45°} = \frac{5\,\text{mm}}{\cos 45°} = 7.071\,\text{mm}$$

Using Eq. 5.11, you can calculate

$$\sigma = \frac{4.24(50{,}000)\,\text{Nm}\left(\dfrac{1000\,\text{mm}}{\text{m}}\right)}{(7.071)\,\text{mm}|300^2 + 3(120)(300 + 7.071|\text{mm}^2} = 149.5\,\text{MPa}$$

Hicks, p. 87–88 |4|, has a more comprehensive solution. However, the preceding equation is much simpler and very accurate.

TABLE 5.1 Wire and Sheet-Metal Gauges (Diameters and thickness values are in decimals of an inch)

Gauge No.	American wire gauge, or Brown and Sharpe (for copper wire)	Steel wire gauge, or Washburn and Moen or Roebling (for steel wire)	Birmingham wire gauge (B.W.G.) (for steel wire or sheets)	Stubs steel wire gauge	U.S. standard gauge for sheet metal (iron and steel) 480 lb per cu ft	AISI inch equivalent for U.S. steel sheet thickness
0000000		0.4900			0.500	
000000		0.4615			0.469	
00000		0.4305			0.438	
0000	0.460	0.3938	0.464		0.406	
000	0.410	0.3625	0.425		0.375	
00	0.365	0.3310	0.380		0.344	
0	0.325	0.3065	0.340		0.312	
1	0.289	0.2830	0.300	0.227	0.281	
2	0.258	0.2625	0.284	0.219	0.266	
3	0.229	0.2437	0.259	0.212	0.250	0.2391
4	0.204	0.2253	0.238	0.207	0.234	0.2242
5	0.182	0.2070	0.220	0.204	0.219	0.2092
6	0.162	0.1920	0.203	0.201	0.203	0.1943
7	0.144	0.1770	0.180	0.199	0.188	0.1793

(continued)

TABLE 5.1 Continued

Gauge No.	American wire gauge, or Brown and Sharpe (for copper wire)	Steel wire gauge, or Washburn and Moen or Roebling (for steel wire)	Birmingham wire gauge (B.W.G.) (for steel wire or sheets)	Stubs steel wire gauge	U.S. standard gauge for sheet metal (iron and steel) 480 lb per cu ft	AISI inch equivalent for U.S. steel sheet thickness
8	0.128	0.1620	0.165	0.197	0.172	0.1644
9	0.114	0.1483	0.148	0.194	0.156	0.1495
10	0.102	0.1350	0.134	0.191	0.141	0.1345
11	0.091	0.1205	0.120	0.188	0.125	0.1196
12	0.081	0.1055	0.109	0.185	0.109	0.1046
13	0.072	0.0915	0.095	0.182	0.094	0.0897
14	0.064	0.0800	0.083	0.180	0.078	0.0747
15	0.057	0.0720	0.072	0.178	0.070	0.0673
16	0.051	0.0625	0.065	0.175	0.062	0.0598
17	0.045	0.0540	0.058	0.172	0.056	0.0538
18	0.040	0.0475	0.049	0.168	0.050	0.0478
19	0.036	0.0410	0.042	0.164	0.0438	0.0418
20	0.032	0.0348	0.035	0.161	0.0375	0.0359
21	0.0285	0.0317	0.032	0.157	0.0344	0.0329

22	0.0253	0.0286	0.028	0.155	0.0312	0.0299
23	0.0226	0.0258	0.025	0.153	0.0281	0.0269
24	0.0201	0.0230	0.022	0.151	0.0250	0.0239
25	0.0179	0.0204	0.020	0.148	0.0219	0.0209
26	0.0159	0.0181	0.018	0.146	0.0188	0.0179
27	0.0142	0.0173	0.016	0.143	0.0172	0.0164
28	0.0126	0.0162	0.014	0.139	0.0156	0.0149
29	0.0113	0.0150	0.013	0.134	0.0141	0.0135
30	0.0100	0.0140	0.012	0.127	0.0125	0.0120
31	0.0089	0.0132	0.010	0.120	0.0109	0.0105
32	0.0080	0.0128	0.009	0.115	0.0102	0.0097
33	0.0071	0.0118	0.008	0.112	0.0094	0.0090
34	0.0063	0.0104	0.007	0.110	0.0086	0.0082
35	0.0056	0.0095	0.005	0.108	0.0078	0.0075
36	0.0050	0.0090	0.004	0.106	0.0070	0.0067
37	0.0045	0.0085		0.103	0.0066	0.0064
38	0.0040	0.0080		0.101	0.0062	0.0060
39	0.0035	0.0075		0.099		

(continued)

TABLE 5.1 Continued

Gauge No.	American wire gauge, or Brown and Sharpe (for copper wire)	Steel wire gauge, or Washburn and Moen or Roebling (for steel wire)	Birmingham wire gauge (B.W.G.) (for steel wire or sheets)	Stubs steel wire gauge	U.S. standard gauge for sheet metal (iron and steel) 480 lb per cu ft	AISI inch equivalent for U.S. steel sheet thickness
40	0.0031	0.0070		0.097		
41		0.0066		0.095		
42		0.0062		0.092		
43		0.0060		0.088		
44		0.0058		0.085		
45		0.0055		0.081		
46		0.0052		0.079		
47		0.0050		0.077		
48		0.0048		0.075		
49		0.0046		0.072		
50		0.0044		0.069		

INTERNAL EXPANSION JOINTS

Internal expansion joints are used to accommodate differential expansion of vessel components. The most common type of expansion joint used in pressure vessels is bellows expansion joints. These joints are normally thin-walled, with a normal minimum wall of 1/16″ (1.5 mm). When these devices are used, there should be a contingency plan for accessing them when they fail. Most bellows expansion joints have a life span of 10–11 years. If a bellows expansion joint is located in a space with no access, then the shell wall must be cut open, which results in labor-intensive and complex repairs that require a large number of man hours. Access space in the expansion joint enclosure must be designed for personnel to perform maintenance. The access space should also contain enough room to accommodate new repair components, such as a *clam shell*. The clam shell is a bellows expansion joint that fits over an existing joint; it eliminates the removal of an existing bellows expansion joint. It has to be welded by a highly skilled welder, and that person needs space to work.

This clam shell device is fabricated welding bellows provided in multiple sections for installation over an existing element or over a piping system that has internal components such as tube bundles. The welding of clam shells requires the longitudinal seams of the bellows element to be welded manually, thus requiring a welder who has a very high degree of skill. These devices can significantly shorten the shutdown or turnaround time required to repair a damaged bellows expansion joint. Along with the repair, expansion joint components, such as hinges and gimbal boxes, can be installed along with pressure monitoring systems to alert you to future leaks.

Expansion joints bellows may be described in terms of gauge of metal, being that they are usually very thin. This term is not uncommon in welding and general plant repair work. For the reader's reference, Table 5.1 lists wire and sheet metal gauges. The Birmingham wire gauge (B.W.G.) is very common for tubes.

REFERENCES

1. Moody, Fredrick, *Predicting Thermal-Hydraulic Loads on Pressure Vessels*, ASME Continuing Education Course Notes PD382. American Society of Mechanical Engineers, New York, NY, 2007.
2. Roark, Raymond J., *Formulas for Stress and Strain*, 4th edition. McGraw-Hill, Table X Case 60, New York, NY, 1965.
3. Giachino, J.W., Weeks, W., and Johnson, G.S., *Welding Technology*, 2nd edition. American Technical Society, Chicago, Illinois, 1973.
4. Hicks, John, *Welded Joint Design*, 3rd edition. Industrial Press, Inc., New York, NY, 1999.
5. Bednar, *Pressure Deisgn Handbook*, 2nd Edition. Van Nostrand Reinhold Company, NY, NY, 1986.

Chapter 6

Safety Considerations for Lifting and Rigging

Safety cannot be overemphasized in almost all applications, but especially in lifting and rigging of equipment. These principles are valid for whatever is to be lifted, whether it involves pressure vessels and stacks, marine shipyard applications, aerospace use, or any of the many more needs for lifting and rigging.

The purpose of this chapter is to highlight safety issues and, more important, relevant safety standards. Also listed are helpful field reference cards, which are used widely in the field for valuable reference material; they are available on weatherproof pocket cards for easy access.

THE CONCEPT OF A TON WEIGHS HEAVY WHEN LIFTING

The unit of measurement, the *ton*, means different things to different people. In the metric SI, the concept is simple: A metric ton equals 1,000 kilograms. However, in U.S. Customary units, the ton can be tricky. The ton in U.S. Customary units is derived from hundredweights. A *short ton* and *long ton* are equal to 20 hundredweights; however, a short hundredweight is equal to 100 pounds, and a long hundredweight is equal to 112 pounds. Thus, a short ton is

$$(20) \text{ hundredweights} (100) = 2000 \text{ lb}_m$$

The long ton is likewise

$$(20) \text{ hundredweights} (112) = 2240 \text{ lb}_m$$

In lifting and in general practice, the short ton is used in the United States. However, in countries that formerly used the Imperial system of units, the long ton is referred to as a *weight ton* or *gross ton*. The long ton is used for petroleum products such as oil tankers hauling petroleum products. This trend is changing toward the wider use of the metric ton.

The metric ton is sometimes referred to as the *tonne* and is equal to 1,000 kilograms or 2,204.6 lb_m. Lifting and rigging in the United States use the short ton, meaning 2000 lb_m. The short ton is simply called *ton* in the United States, or sometimes *net ton*.

Another example of the advantage of the metric SI system of units is that you do not have to worry about long or short, but simply a metric ton, or tonne. The issue is confusing to the point that purchasing groups apply all units for each pressure vessel or stack shipped. Normally, a chart is made showing the shipping weight of the item in short tons, long tons, and metric tons to avoid confusion.

MAXIMUM CAPACITY OF SLINGS

The angle between a sling and a horizontal plane is very important. This concept is illustrated by the simple double sling shown in Figure 6.1. When looking at this figure, you can make some basic assumptions. One assumption is that the two slings are attached to a common body, or mass. The second assumption is that the center of gravity, or CG (to be discussed later), is equally distant between the two lift points.

The capacity of the two slings is a function of the angle θ. For equilibrium to exist,

$$2F \sin \theta = LOAD \tag{6.1}$$

Solving for the load in each sling, you have

$$F = \frac{LOAD}{2 \sin \theta} \tag{6.2}$$

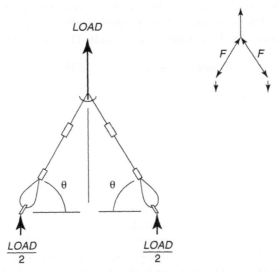

FIGURE 6.1 Load vectors and sling angles. The force vectors are shown in the upper right. The two slings form is known as a two-leg bridle.

Thus, using U.S. Customary units, if $LOAD = 50$ tons and $\theta = 60°$, then the force in each sling is

$$F = \frac{LOAD}{2 \sin \theta} = \frac{50 \text{ tons}}{2 \sin(60)} = 28.87 \text{ tons}$$

For $\theta = 45°$,

$$F = \frac{50 \text{ tons}}{2 \sin(45)} = 35.36 \text{ tons}$$

and for $\theta = 30°$,

$$F = \frac{50 \text{ tons}}{2 \sin(30)} = 50 \text{ tons}$$

This discussion is in conformance with the ANSI/ASME B30.9 and good engineering practice.

Thus, the maximum capacity (ML) of each sling configuration is proportional to the sling angle, θ. With a sling angle of 30°, the force in the sling is $50/28.87 = 1.732$ times the force in the sling using a sling angle of $\theta = 60°$. Thus, using a sling angle of $\theta = 60°$, the sling has 1.732 times the capacity of using a sling angle of $\theta = 30°$. Stated differently, using a sling angle of 30° results in a force in the sling which is 1.732 times the force in a sling if one were using a sling angle of 60°.

Sling manufacturers use the concept of ML in maximum capacity charts. In Figure 6.1 using $\theta = 30°$, the load in each sling is equal to the $LOAD$. Hence, for two slings with $\theta = 60°$, a sling manufactured for taking a tensile force equal to the $LOAD$, the maximum capacity for the two slings, ML, would be $1.732 \times LOAD$, or

$$ML = (2 \sin \theta) \, LOAD \tag{6.3}$$

Thus, for $\theta = 60°$,

$$ML = [2 \sin(60)] \, LOAD = 1.732 \, LOAD \tag{6.4}$$

Similarly, for

$$\theta = 45°, \quad ML = [2 \sin(45)] \, LOAD = 1.4142 \, LOAD \tag{6.5}$$

$$\theta = 30°, \quad ML = [2 \sin(30)] \, LOAD = LOAD \tag{6.6}$$

It is common practice to make the sling angle, θ, equal to or greater than 45°, preferably 60°.

Example 6.1 The Kilo Newton Problem

For this example, suppose you are in Australia and have to lift an American-built tank that weighs $2500\,lb_m$. The crane from Germany you have available is rated at $20\,KN$. The metric SI is the only acceptable system of units in Australia, so you must talk, speak, and calculate in metric SI.

From Chapter 1 remember the following:

$$F = (1.0\,Kg) \left(\frac{9.807\,\dfrac{m}{sec^2}}{1\,\dfrac{Kg - m}{N - sec^2}} \right) = 9.807\,N \qquad (1.3)$$

The German crane could safely lift

$$20\,KN = 20,000\,N$$

Using Eq. 1.3, you have

$$\text{Mass}\,(Kg) = F^* \left(\frac{g_c}{g} \right) = 20,000\,N * \left(\frac{1\,\dfrac{Kg - m}{N - sec^2}}{9.807\,\dfrac{m}{sec^2}} \right) = 2039.4\,Kg$$

So the crane is rated at roughly $2039\,Kg$. Using the U.S. Customary system of units, you calculate $2039.4\,Kg = 4496.1\,lb_m > 2500\,lb_m$, so the crane is adequate. This is a factor of safety (FOS) of

$$FOS = \frac{4496.1}{2500} = 1.8$$

Now with a sling angle of $60°$, the sling capacity configuration with two slings would be from Eq. 6.4:

$$ML = 1.732\,LOAD = 1.732(2500) = 4330\,lb_m = 1964.05\,Kg$$

You now know that the maximum capacity of both slings exceeds $2500\,lb_m$ ($1134\,Kg$) and is thus adequate. Using a typical sling manufacturer chart, shown in Table 6.1, you can select slings with an angle of $60°$ rated at, say, $1500\,Kg$, and a diameter of the sling chain of $8\,mm$.

TABLE 6.1 A Typical Sling Manufacturer Chart (Courtesy of Slingmax, Inc.)

Chain Size Inches	Single Branch Sling 90 degree Loading	Double Sling			Triple and Quadruple Sling		
		60 degree	45 degree	30 degree	60 degree	45 degree	30 degree
9/32	3,500	6,100	4,900	3,500	9,100	7,400	5,200
3/8	7,100	12,300	10,000	7,100	18,400	15,100	10,600
1/2	12,000	20,800	17,000	12,000	31,200	25,500	18,000
5/8	18,100	31,300	25,800	18,100	47,000	38,400	27,100
3/4	28,300	49,000	40,000	28,300	73,500	60,000	42,400
7/8	34,200	59,200	48,400	34,200	88,900	72,500	51,300
1	47,700	82,600	67,400	47,700	123,900	101,200	71,500
1-1/4	72,300	125,200	102,200	72,300	187,800	153,400	108,400
1-1/2	80,000	138,600	113,100	80,000	—	—	—

BRIDLES AND CENTER OF GRAVITY (CG)

Figure 6.1, earlier in the chapter, shows a two-leg bridle. When you are using bridles, lifting lugs, and trunnions, the center of gravity, or CG, is all important on how the loads are distributed. This concept is illustrated in more detail in Chapter 7, where we demonstrate lifting analyses.

Suppose you have a four-leg bridle. The loads in the slings vary because the slings are almost always unequal in length. The loads are statically indeterminate, meaning that the true load in each sling cannot be mathematically solved. In reality, the load is carried by two slings, while the other two act to balance the load. To solve this problem, you must size the bridle such that just two legs carry the full load, or you must use a spreader.

LIFT CATEGORIES

There are four lift categories described as follows:

- *Light Lift:* Any lift where the payload lift is 10 tons or less.
- *Medium Lift:* Any lift where the payload weight is over 10 tons but less than 50 tons.
- *Heavy Lift:* Any lift where the payload lift is 50 tons or greater.
- *Critical Lift:* Any lift that exceeds 90% of the crane's chart capacity; or any multiple-crane lift where either crane exceeds 75% of the crane's load capacity or requires one or both cranes to change locations during the lifting operation; or any lift over operating or occupied facilities, operating process pipe racks, or near power lines; any lift involving complex rigging arrangement or that requires specialty rigging; also any lifting operation involving sensitive or risk to costly equipment.

In Chapter 7 we will discuss lifting hydrocrackers that have over a 1,000 metric ton lift weight. These cases are critical lifts because of the payload, and consequences of failure are prohibitive.

PREPARING FOR THE LIFT

It is always prudent to check with equipment manufacturers on how to lift equipment they designed and fabricated. In Chapter 3, the example of guy cables for flare stacks was a real scenario in Saudi Arabia. The flare tip arrived and had no lifting devices on it. A project engineer asked me to design and install lifting lugs on the flare tip. Upon investigation, I discovered the flare tip was made of Alloy 600 and had very low specified minimum yield strength, meaning that the lifting lugs would have to be massive. I advised the project engineer to contact the flare tip manufacturer on how to lift the tip to the top of the flare tip. The flare tip manufacturer responded that it purposely did not install lifting lugs on the flare tip and that we should use plastic slings wrapped around it for lifting! The riggers used the plastic rope and successfully lifted and installed the flare tip without any lifting lugs.

Another problem that lifting lugs could have caused is unequal heat distribution around the ring during flare. This is a classic example of making sure you lift the equipment as it is intended. **Note:** the same is true for the rigging. Be sure to refer to hoist and rigging equipment manufacturers' specifications for proper applications and limitations.

AMERICAN NATIONAL STANDARD INSTITUTE (ANSI) SAFETY CODES

ANSI standards provide comprehensive guidelines for the variety of equipment and work operations for rigging work. Many of these standards are enforced by the Occupational Safety and Health Administration (OSHA), among other safety regulations. These standards are as follows:

- ANSI B30.1 JACKS: Addressed in this standard are safety requirements for construction, installation, operation, inspection, and maintenance of screw, ratchet, lever, and hydraulic jacks. Minimum inspection requirements are included before jacks are employed in use.
- ANSI B30.20 Overhead and Gantry Cranes
- ANSI B30.3 Hammerhead Tower Cranes
- ANSI B30.4 Portal, Tower, and Pillar Cranes
- ANSI B30.6 Derricks
- ANSI B30.8 Floating Cranes and Floating Derricks
- ANSI B30.11 Monorail Systems and Underhung Cranes
- ANSI B30.13 Controlled Mechanical Storage Cranes
- ANSI B30.14 Side Boom Tractors
- ANSI B30.17 Overhead and Gantry Cranes (Top Running Bridge, Single Girder, Underhung Hoist)
- ANSI B30.18 Stacker Cranes
- ANSI B30.22 Articulating Boom Cranes
- ANSI B30.24 Container Cranes
- ANSI B30.25 Material Handling Hybrid Cranes: Safety requirements for various kinds of cranes are established in this standard. Provided are frequent and periodic inspection requirements, operator qualifications, and standard hand signals. Emphasized is the importance of utilizing the right standard for the right kind of crane.
- ANSI B30.7 Base Mounted Drum Hoists
- ANSI B30.16 Overhead Hoists (Underhung)
- ANSI B30.21 Manually Lever Operated Hoists: Provisions for detailed requirements for hoists which are frequently used in construction rigging work are covered in this standard.

- ANSI B30.9 Slings: Provisions are made for a comprehensive set of safety standards for the use and periodic inspection of alloy steel chain, wire rope, metal mesh, natural and synthetic fiber rope, and synthetic webbing (nylon, polyester, and polypropylene). The rigging personnel must have a good working knowledge of this standard to effectively design and use slings in construction work operations.
- ANSI B30.10 Hooks
- ANSI B30.12 Handling Loads from Suspended Rotorcraft (Helicopters)
- ANSI B30.19 Cableways
- ANSI B30.20 Below-the-Hook Lifting Devices: In this standard are provisions for lifting devices such as lifting beams (spreader beams), edge grip sheet clamps, and plate clamps. The requirements for the design, fabrication, inspection, and use of lifting beams.
- ANSI B30.23 Personnel Lifting
- ANSI B56.1 Lift and High Lift Trucks (Forklifts)
- ANSI B56.5 Guided Industrial Vehicles
- ANSI B56.6 Rough Terrain Forklift Trucks
- ANSI B56.7 Industrial Crane Trucks
- ANSI B56.8 Personnel and Burden Carriers
- ANSI B56.9 Operator Controlled Industrial Tow Tractors
- ANSI N45.15 Hoisting, Rigging, and Transporting of Items at Nuclear Plants.

HELPFUL REFERENCES FOR RIGGING

The following references list many helpful rigging tips:

Rigger's Pocket Guide, by Construction Safety Association of Ontario, 21 Voyager Court South, Etobicoke, Ontario, Canada M9W 5M7, 1-800-781-2726.

Journeyman Rigger's Reference Card, by Parnell Services Group Training & Inspection Resource Center, PO Box 1660, Woodland, WA 98674, in U.S. 1-888-567-8472.

Master Rigger's Reference Card, by Parnell Services Group Training & Inspection Resource Center, PO Box 1660 Woodland, WA 98674, in U.S. 1-888-567-8472.

Lineman Rigger's Reference Card, by Parnell Services Group Training & Inspection Resource Center, PO Box 1660, Woodland, WA 98674, in U.S. 1-888-567-8472.

Rigging Gear Inspection Card (Per ASME B30.9 & 29 CFR 1910.184), by Parnell Services Group Training & Inspection Resource Center, PO Box 1660, Woodland, WA 98674, in U.S. 1-888-567-8472.

Equipment Operator's Card, by Parnell Services Group Training & Inspection Resource Center, PO Box 1660, Woodland, WA 98674, in U.S. 1-888-567-8472. This handy card list components that need to be checked for mobile cranes and boom trucks, overhead cranes, forklifts, tractor rigs/large trucks, load securement, graders, backhoes, dozers, yard tractors, and snow vehicles.

Mobile Crane Operator Reference Card, by Parnell Services Group Training & Inspection Resource Center, PO Box 1660, Woodland, WA 98674, in U.S. 1-888-567-8472. A handy field reference card for planning the lift, from checking out the crane to hitch types and wire rope capacities.

Chapter 7

Lifting and Tailing Devices

This chapter describes the lifting of pressure vessels and stacks in the field. The perspective is from the field, and not design, viewpoint. This information should be valuable to engineering designers, but, as mentioned, the information is more from a field perspective. Unfortunately, many engineering designers do not work in the field and, consequently, are not exposed to operational and field problems. This chapter therefore should serve as a source of information for both field and engineering design offices.

Lifting and tailing devices are extremely important because their failure can result in the loss of property and lives. The responsibility for lifting systems is very explicit, making the parties responsible accountable.

This chapter is partially based on the work of Duerr [1] and Bragassa [2]. Duerr's work is validated with laboratory tests by various investigators. What we are first concerned with are the modes of failures and how to prevent them. First, we will address tail and lifting lugs for lifting pressure vessels and stacks. Later, we will address lifting trunnions, which are different lifting devices than lugs.

Lifting (and tailing) lugs are pinned connections consisting of a pin extended through the lug hole connected to a shackle or a link-pin arrangement. We will discuss the latter arrangement later in this chapter. The stress distribution in a lug pin arrangement is very complex; consequently, the design code requirements are empirical. However, these empirical relationships have worked over many decades, and the stress profile of the pin connection is understood well enough for general application.

The lug geometry is shown in Figure 7.1.

Using Figure 7.1 as a starting point, we will discuss the five basic mechanisms of failure and expand on the others. The five basic mechanisms of failure of a lug plate, as outlined by Duerr [1], are as follows:

1. Tension at net section
2. Hoop tension (splitting failure beyond hole)
3. Double plane shear failure
4. Failure by out-of-plane instability (dishing)
5. Bearing failure

These failure modes are graphically shown in Figure 7.2(a) and 2(b). These modes of failure act independently and do not occur at the same time.

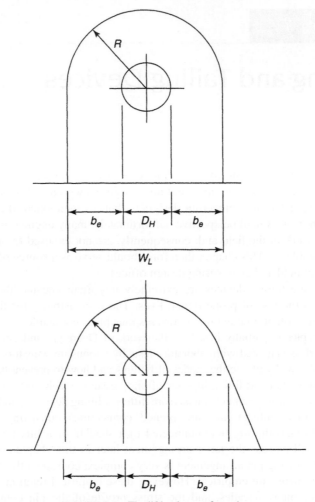

FIGURE 7.1   The lifting lug configuration.

IMPACT FACTOR

In the following discussion the use of impact factors is performed in the examples, whether mentioned or not. The impact factor, also known as the dynamic load factor, is used to consider crane slippage, wind gusts, or any other factor resulting in an increase load. The impact factor can vary from 1.25 to 2.0. Typically, 1.5 is used. This factor varies with each company. For the heavy vessels mentioned in the examples in this chapter, the client used an impact factor of 1.35. This factor is not applied to below-the-hook lifting devices, such as shackles. Shackles are proof tested to 1.33 to 2.2, the working load limit, depending on the capacity (this may vary with the manufacturer). Thus shackles have a built-in factor of safety making the application of an impact factor unnecessary.

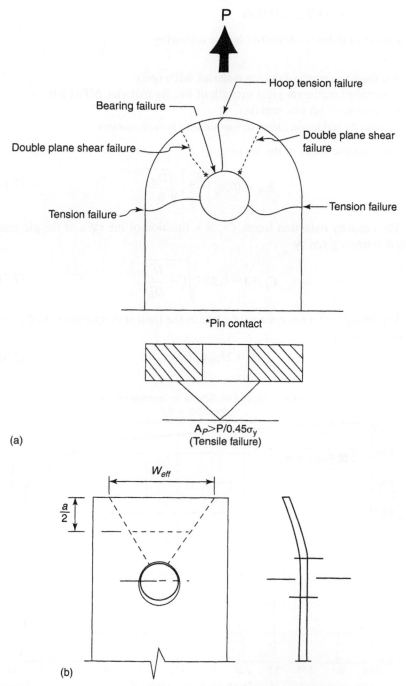

FIGURE 7.2(a) The modes of failure of a lifting lug (dishing is shown in Figure 7.2(b))
(b) Four of the five modes of failure of a lifting lug. In dishing, the diagram at the top right is a side view of a lug that is deformed, or dished.

TENSION AT NET SECTION

This mode of failure is described by the following:

Let
F_u = ultimate strength of the lug material, MPa (psi)
F_y = specified minimum yield strength of the lug material, MPa (psi)
D_p = diameter of lift pin, mm (in)
C_r = capacity reduction factor of the pin and hole diameters

The effective width of the lug is

$$b_{eff} = 0.6b_e \left(\frac{F_u}{F_y}\right)\sqrt{\frac{D_H}{b_e}} \qquad (7.1)$$

The capacity reduction factor, Cr, is a function of the ratio of the pin and hole diameters given by

$$C_r = 1 - 0.275\sqrt{1 - \frac{D_p^2}{D_H^2}} \qquad (7.2)$$

The strength of a pin-connected plate in the limit state of tension in the net section is given by

$$P_n = 2b_{eff}C_r b_{eff} tF_u \qquad (7.3)$$

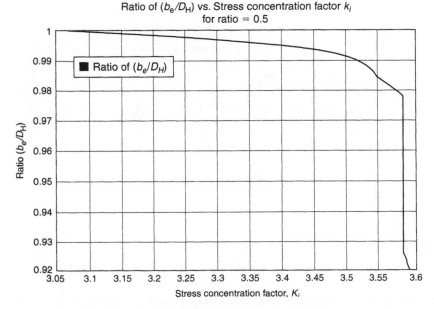

FIGURE 7.3 Ratio of the ratio b_e/D_H to stress concentration factor, K_i.

If $P_n > P$, the applied load, then the lug plate is satisfactory for tension at the net section.

The pin clearance in the lug hole as a function of the stress concentration factor has been determined in lab tests. Figure 7.3 shows the pin-to-lug hole ratio plotted against the stress concentration factor for $b_e/D_H = 0.5$.

Figure 7.4 shows a plot of the capacity reduction factor, Cr, plotted against the ratio D_p/D_H. In practice the lug hole diameter is 3 mm (1/8"). From the stress viewpoint, as shown in Figure 2.2 and Figure 7.3, the closer D_p is to D_H, the lower the stress. However, you must be careful not to specify something that cannot be built. Using a clearance between the pin and lug diameters of 0.8 mm (1/32") is extremely difficult to accomplish. One reason is that, after the vessel is lifted, all it takes is for a pin to deflect a very slight amount, and the pin cannot be removed from the lug hole. Second, when paint is added to the lifting lug, the pin will not fit. When such tight clearances are used, the worst event usually happens. The construction personnel cut a larger (and uneven) hole with a weld torch in the lug plate, resulting in an undesirable situation. The resulting hole leaves a configuration that was not considered in the calculations.

This newly cut hole presents an interesting dilemma when modeling a pin connection in a lug plate hole with the finite element method (FEM). The angle of contact between the pin and lug plate is very small—5° or less, depending on the ratio of the diameter of the pin to the diameter of the lug hole. Point contacts are very difficult to model in FEM; the reason is that, if forces are not distributed over several elements, the resulting stress can be enormous and unrealistic. When the pin and lug come into contact, localized yielding occurs in the pin and lug plate.

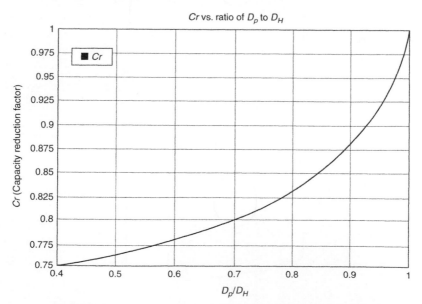

FIGURE 7.4 Capacity reduction factor, Cr, versus the ratio of D_p/D_H.

The closed-form formulations have proven adequate throughout many years. The laboratory tests regarding this issue provide invaluable data for the use of closed form solutions. The reason for designers specifying a very low Dp/D_H ratio of 0.8 mm (1/32") is to provide a greater area of contact. When localized deformation results, which it inevitably will, such tight tolerances are not only unnecessary, but impossible to execute in the field. Thus, 3 mm (1.8") should be adequate for the difference between the pin diameter and lug hole diameter.

HOOP TENSION—SPLITTING FAILURE BEYOND HOLE

The hoop tension phenomenon is perhaps the most likely mode of failure. This mode occurs when the lift force acts in tension in the hoop direction, as shown in Figure 7.5. This mode of failure is the tensile force acting on the area of the lug from the top of the lug hole to the lug edge in one plane, and the lug

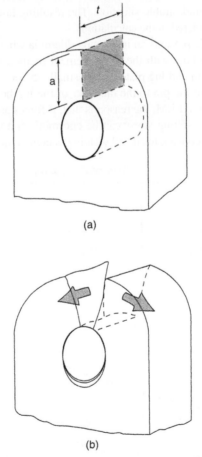

(a)

(b)

FIGURE 7.5   Hoop tension forces splitting apart a lifting lug.

thickness in the other plane. This is what is meant by the "hoop" direction. The hoop tensile force that tends to pull the lug apart is resisted by a direct shear through a single plane. In the *AISC Manual of Steel Construction* [3], Chapter D, Paragraph D3.1, states that the allowable stress on the net area of the pin hole for pin-connected members is $0.45Fy$. Sometimes this mode of failure is referred to as "tensile splitting." The net area, A, is $(a)(t)$, as shown in Figure 7.5(a). Figure 7.5(b) shows the actual failure where the lug splits. Using the AISC criterion, you can use the equation for the hoop tension as follows:

$$t_r = \frac{P}{0.45 F_y a} \qquad (7.4)$$

where
$a = R - D_H/2$ shown in Figure 7.5(a), mm (in)
F_y = specified minimum yield strength of the lug material, MPa (psi)
P = tensile load on the lifting lug, N, lb$_f$
t_r = required lug thickness, mm (in)

The strength of a pin-connected plate for the net area above the hole, A, is given by Duerr [1] as follows:

$$P_b = C_r F_u \left[1.13\, a + \frac{0.92b}{D_H} \right] t \qquad (7.5)$$

where all the terms in the equation are as defined previously. Now if $P_b \geq P$ and $t \geq t_r$, where t = the actual lug thickness, then the lug is satisfactory for hoop tension.

Hoop tension is quite often the governing mode.

DOUBLE PLANE SHEAR FAILURE

The double plane shear mode of failure is graphically shown in Figure 7.6. The parameters associated with this failure are shown in Figure 7.7.

The region beyond the hole is that of shear on two planes which are parallel to each other and the vector of the acting force on the lug, shown in Figure 7.6. The locations of the shear planes are defined by the angle, ϕ, shown in Figure 7.7. The shear plane consists of two vertical lines from the point defined by ϕ. The ultimate strength of the material decreases as the clearance between the pin and lug hole increases, provided all other dimensions remain constant. Duerr [1] records lab tests where a relationship for ϕ was developed to account for this clearance.

The computation of the shear strength of the plate requires knowledge of the ultimate shear strength of the material, F_{us}. This property is not available

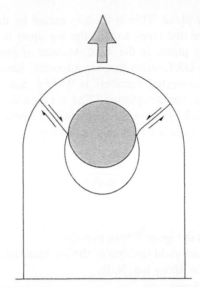

FIGURE 7.6   Double plane shear mode of failure.

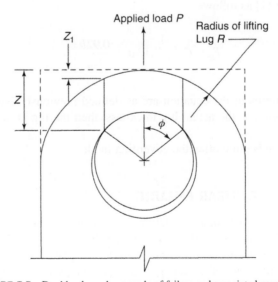

FIGURE 7.7   Double plane shear mode of failure and associated parameters.

and usually must be determined by empirical tests. For purposes of this example, to be conservative, you have

$$F_{us} = 0.7F_u \qquad (7.6)$$

The angle ϕ is defined as follows:

$$\phi = 55 \frac{D_P}{D_H} \tag{7.7}$$

The length of the shear plane, shown in Figure 7.7, is as follows:

$$Z = a + \frac{D_P}{2}(1 - \cos \phi) \tag{7.8}$$

For a lug with an outer surface that is a circular arc, as in Figure 7.7, the dimension Z_1 is calculated as follows:

$$Z_1 = R - \sqrt{R^2 - \left(\frac{D_P}{2}\sin\phi\right)^2} \tag{7.9}$$

For a rectangular or square lug, where the outer surface is a straight line, then $Z_1 = 0$.

For the strength of a pin-connected plate in the limit state of a double plane shear for a lug with a circular edge,

$$P_s = 2(Z - Z_1)F_{us} \tag{7.10}$$

For a rectangular plate with a straight top edge,

$$P_s = 2ZF_{us} \tag{7.11}$$

Now, continuing with the circular edge lug plate, the area of shear is

$$A_{shear} = 2(Z - Z_1)t \tag{7.12}$$

The shear stress in the lug plate is

$$\tau = \frac{P_s}{A_{shear}} \tag{7.13}$$

The acceptance criteria for double shear are as follows:

If $\tau \le 0.4F_y$ then the lug is satisfactory for double shear.
If $\tau > 0.4F_y$ then the lug is not satisfactory for double shear.

OUT-OF-PLANE INSTABILITY (DISHING) FAILURE

The out-of-plane instability mode of failure occurs with slender pin-connected plates that fail by out-of-plane buckling. This mode is demonstrated in

Figure 7.2(b). The plate above the pin is analogous to a cantilevered beam. The critical buckling stress can be expressed in terms of a slenderness ratio KL/r, where L is equal to the plate dimension a in Figure 7.5 and r is the radius of gyration of the plate through the thickness direction $(t/\sqrt{12})$. Tests indicate that the length of the cantilever is not necessarily equal to a. Plates that are wide—with a relatively larger value of b_e—are seen to provide less support to the area in compression above the hole, resulting in a larger effective length. This means that the wider the spread of the centroids on each side of the hole, the more likely the plate will buckle. This phenomenon is accounted for with the effective length factor, K, as follows:

$$K = 2\sqrt{\frac{b_e}{a}} \qquad (7.14)$$

The lug plate can either fail elastically or inelastically. Plates in which the following is true will fail inelastically:

$$C_c = \sqrt{\frac{2\pi^2 E}{F_y}} < \frac{KL}{r} \qquad (7.15)$$

The inelastic critical buckling stress is given as follows:

$$F_{cr} = \left[\frac{1 - \dfrac{(KL/_r)^2}{2C_c^2}}{1 - v^2}\right] F_y \qquad (7.16)$$

where v = Poisson's ratio (= 0.3 for steel).

The elastic critical buckling stress is given as follows:

$$F_{cr} = \frac{\pi^2 E}{(KL/_r)^2 (1 - v^2)} \qquad (7.17)$$

Referring to Figure 7.2(b), the critical buckling stress acts on an effective area of the plate, equal to W_{eff}/t. W_{eff} is an effective width shown in the figure and the lesser of the values given by Eq. 7.16 or Eq. 7.17. The following is analogous to the effective width model used for some edge-loaded plate buckling problems:

$$W_{eff} = D_p + a \qquad (7.18)$$

The following is an upper limit value determined from test data:

$$W_{\text{eff}} = D_H + 1.25b_e \tag{7.19}$$

The strength of a pin-connected plate in the limit state of dishing is as follows:

$$P_d = W_{\text{eff}}tF_{cr} \tag{7.20}$$

where F_{cr} is given in Eq. 7.16 and Eq. 7.17 and W_{eff} is given in Eq. 7.18 and Eq. 7.19.

The last acceptance criterion is that no fishing occurs in plates for which the proportions are defined as follows:

$$\text{DISHRATIO} = \frac{(a)(D_H)}{t(D_p)} \tag{7.21}$$

If $\text{DISHRATIO} < \sqrt{\dfrac{E}{F_y}}$ then the acceptance criterion is met. If the DISHRATIO is equal to or greater than $\sqrt{\dfrac{E}{F_y}}$, then the acceptance criterion is not met.

Duerr's important work [1] is the basis for the ASME BTH-1-2005 Standard, *Design of Below-the-Hook Lifting Devices* [4] and the ASME B30.20-2006, *Below-the-Hook Lifting Devices* [5].

BEARING FAILURE

Bearing failure is defined by the following:

$$\sigma_B = \frac{P}{D_p(T_L + T_{DP})} \tag{7.22}$$

where
σ_B = Bearing stress, MPa (psi)
P = Load on lifting lug, N (lb$_f$)
D_p = Pin diameter, mm (in)
T_L = Thickness of lifting lug, mm (in)
T_{DP} = Thickness of doubler plate, mm (in), $T_{DP} = 0$ with no doubler plates

The term "doubler plate" is often referred to as "ears" or "collar plates." We will use the term "doubler plate."

The allowable stress criterion (acceptance criterion) for bearing stress is $\sigma_B \le 0.9F_y$.

Validation Tests for Bearing Failure

Lab tests to confirm equations used in the bearing deformation of lifting lugs were chronicled by Duerr [1]. We will present just a summary of the results for practicing engineers.

The stiffness coefficient, K_{br}, attributable to shear deformation beyond the lug hole, is as follows:

$$K_{br} = 120tF_y \left(\frac{D_p}{25.4} \right)^{0.8}$$ (7.23)

When you use U.S. Customary Units (USCU), Eq. 7.23 becomes

$$K_{br} = 120tF_y D_p{}^{0.8}$$ (7.24)

The equations for K_{br} are based on the model shown in Figure 7.8. Referring to Figure 7.8, the pin bearing area is given by

$$A_p = \sin(\alpha_1)D_p t$$ (7.25)

where

$$\alpha_1 = \text{Arc} \cos \left(\frac{0.25D_H^2 - 0.25D_p^2 - \Delta^2}{D_p \Delta} \right)$$ (7.26)

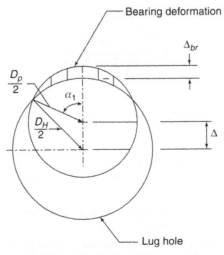

FIGURE 7.8 Bearing stiffness model.

$$\Delta = \frac{D_H}{2} - \frac{D_p}{2} + \Delta_{br} \tag{7.27}$$

Lab tests indicate that a $\Delta_{br} = 0.1\,\text{mm}$ is a good initial assumption, and that plate deformations are linear up to the value of $\Delta_{br} = 0.25\,\text{mm}$. The pin load,

$$P = A_p F_y, \tag{7.28}$$

and the stiffness,

$$K_{br} = \frac{P}{\Delta_{br}}, \tag{7.29}$$

result in the following:

$$K_{br} = 10C \sin(\alpha_1) D_p t F_y \tag{7.30}$$

where $C = 1.0$ in SI units and 25.4 in USCU. In USCU when $10C\sin(\alpha_1) = 120$ for the range of specimen dimensions reported in lab tests, Eq. 7.30 gives results similar to Eq. 7.24. You can use the stiffness values given by Eq. 7.30 to compute local bearing deformations up to maximum value $\Delta_{br} = 0.25\,\text{mm}$.

PIN HOLE DOUBLER PLATES

If you want to increase the bearing area, it is common practice to reinforce lifting lugs with doubler plates shown in Figure 7.9. These plates were described in the

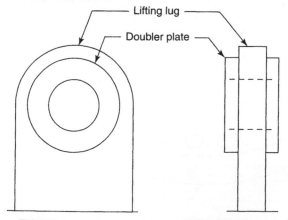

FIGURE 7.9 Lifting lug reinforced with doubler plates.

section on Bearing Failure. Calculation of the strength of these plates is done by calculating the strength of each plate—each doubler plate and the lug plate—and summing the values. This approach yields good agreement with lab tests.

Another approach is to assume the applied load is distributed between the pin and the plates as uniform bearing. The use of doubler plates is ubiquitous. However, with heavy lifts where the lug plate thickness exceeds 100 mm (approximately 4 inches), lifting lugs without doubler plates have successfully been used on many occasions.

Example 7.1 Evaluating a Lifting Lug for Five Modes of Failure

To demonstrate the modes of failure for lifting lugs described by Duerr[1], we use the following example that was successfully used in practice recently. The lug has a maximum tensile load, P, of 10,000,000 Newtons. The lug is shown in Figure 7.10. This lug has a flanged lug design that is welded to a cover plate, which is bolted to the top nozzle on a hemispherical head of a hydrocracker. The vessel has 194 mm of wall made of 2-1/4 Cr-1 Mo metal with 321 austenitic stainless steel lining. Typically, the top head in the center of a hemispherical head is one of the most robust components of a vessel.

The algorithm of the lifting lugs is shown in Figures 7.11(a) and (b).

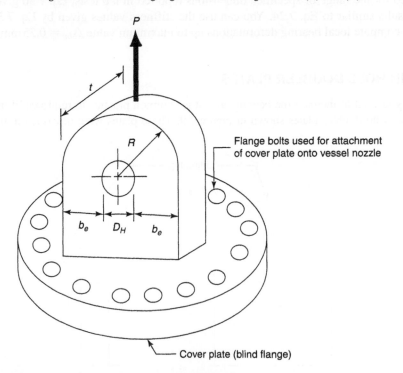

FIGURE 7.10   The flange lifting lug used in actual practice and demonstrated in Example 7.1. Here and in the example, note that D_h is the same as D_H used in the previous discussion.

Rules Sheet:

;TENSION IN NET SECTION

$$beff = 0.6 \cdot be \cdot \left[\frac{Fu}{Fy} \right] \cdot \sqrt{\frac{Dh}{be}}$$

$$Cr = 1 - 0.275 \cdot \sqrt{1 - \frac{Dp^2}{Dh^2}}$$

$Pn = 2 \cdot Cr \cdot beff \cdot t \cdot Fu$

If $Pn > P$ then $Pn = OKT$

If $Pn = P$ then $Pn = OKT$

If $Pn < P$ then $Pn = NOTOKT$

;HOOP TENSION

$$Pb = Cr \cdot Fu \cdot \left[1.13 \cdot a + \frac{0.92 \cdot be}{1 + \frac{be}{Dh}} \right] \cdot t$$

If $Pb > P$ then $Pb = OKHT$

If $Pb = P$ then $Pb = OKHT$

If $Pb < P$ then $Pb = NOTOKHT$

$$HT = \frac{P}{a \cdot t}$$

If $HT < 0.45 \cdot Fy$ then $HT = OK$

If $HT = 0.45 \cdot Fy$ then $HT = OK$

If $HT > 0.45 \cdot Fy$ then $HT = NOTOK$

;DOUBLE PLANE SHEAR FAILURE

$$\varphi = 55 \cdot \left[\frac{Dp}{Dh} \right]$$

$$Z \cdot a + \left[\frac{Dp}{2} \right] \cdot (1 - COS(\varphi))$$

$$Z1 = R - \sqrt{R^2 - \left[\left[\frac{Dp}{2} \right] \cdot SIN(\varphi) \right]^2}$$

$Fus = 0.7 \cdot Fu$

$Ps = 2 \cdot (Z - Z1) \cdot Fus$

$Ashear = 2 \cdot (Z - Z1) \cdot t$

$$\tau = \frac{Ps}{Ashear}$$

If $\tau < 0.4 \cdot Fy$ then $\tau = OKS$

If $\tau = 0.4 \cdot Fy$ then $\tau = OKS$

If $\tau > 0.4 \cdot Fy$ then $\tau = NOTOKS$

FIGURE 7.11(a) Equation sheet for Duerr acceptance criteria.

;FAILURE BY OUT-OF-PLANE INSTABILITY (DISHING)

$$K = 2 \cdot \sqrt{\frac{be}{a}}$$

$$\pi = 3.1416$$

$$r = \frac{t}{\sqrt{12}}$$

$$Cc = \sqrt{\frac{2 \cdot \pi^2 \cdot E}{Fy}}$$

$$RATIO = \frac{K \cdot L}{r}$$

$$L = a$$

If Cc > RATIO then RATIO = OK1

If Cc = RATIO then RATIO = OK1

If Cc < RATIO then RATIO = NOTOK1

$$u = 0.3$$

$$Fcrie = \left[\frac{1 - \left[\dfrac{K \cdot L}{r} \right]^2}{\dfrac{2 \cdot Cc^2}{1 - u^2}} \right]$$

$$Fcre = \frac{\pi^2 \cdot E}{\left[\dfrac{K \cdot L}{r} \right]^2 \cdot \left[1 - u^2 \right]}$$

$$W_{eff} = MIN((D_p + a), (D_H + 1.25 \cdot be))$$

$$Pd1 = W_{eff} \cdot t \cdot Fcrie$$

$$Pd2 = W_{eff} \cdot t \cdot Fcre$$

$$DISHRATIO = \frac{a \cdot D_H}{t \cdot Dp}$$

if $DISHRATIO < 0.19 \cdot \sqrt{\frac{E}{Fy}}$ then DISHRATIO = OK2

if $DISHRATIO = 0.19 \cdot \sqrt{\frac{E}{Fy}}$ then DISHRATIO = NOTOK2

if $DISHRATIO > 0.19 \cdot \sqrt{\frac{E}{Fy}}$ then DISHRATIO = NOTOK2

;BEARING FAILURE

$$\sigma B = \frac{P}{Dp \cdot (t + tDP)}$$

If $\sigma B < 0.9 \cdot Fy$ then σB = OK3

If $\sigma B = 0.9 \cdot Fy$ then σB = OK3

If $\sigma B > 0.9 \cdot Fy$ then σB = NOTOK3

FIGURE 7.11(a)   (continued)

Variables sheet Input	Name	Output	Unit	Comment
				TENSION IN NET SECTION
	beff	304.680828	mm	Effective width of lug
245	be		mm	Horizontal distance from hole centerline from edge of hole to edge of lug
482.76	Fu		MPa	Ultimate strength of lug material
262	Fy		MPa	Specified minimum yield strength of lug material
310	Dh		mm	Diameter of hole in lug
	Cr	0.955965674		Function of ratio of ratio of pin and hole diameters
306	Dp		mm	Diameter of lift pin
	Pn	104051998		Strength of a pin-connected plate for tension in the net section, N
370	t		mm	Thickness of lug plate
	OKT	104051998		If this space is filled then tension in the net section is acceptable
	NOTOKT			If this space is filled then tension in the net section is NOT acceptable
				HOOP TENSION (SPLITTING FAILURE BEYOND HOLE)
	Pb	68771718.8		Strength of a pin-connected plate for net area on top of hole, N
	OKHT	68771718.8		If this space is filled then Hoop Tension is acceptable
	NOTOKHT			If this space is filled then the Hoop Tension is NOT acceptable
245	a		mm	Vertical distance from top edge of lug hole to edge of lug plate
	HT	110.314396	MPa	Hoop Tensile Stress
1E7	P			Maximum tensile load on lifting lug, N
	OK	110.314396		If this space is filled then Hoop Tensile stress is acceptable
	NOTOK			If this space is filled then Hoop Tensile stress is NOT acceptable
				DOUBLE PLANE SHEAR FAILURE
	φ	54.2903226		Angle from lug centerline to shear plane, radians
	Z	495.10066	mm	Vertical distance of shear plane for rectangular lug
	Z1	17.8749705	mm	Vertical distance of shear plane in lug with radius of curvature
400	R		mm	Radius from lug hole centerline to curved edge of lug
	Fus	337.932	MPa	Ultimate shear strength
	Ps	322539.664		Strength of pin-connected plate in double shear, N

FIGURE 7.11(b) Variable sheet showing results and answers for the lifting lug.

Variables sheet

Input	Name	Output	Unit	Comment
	Ashear	353147.01		Area of double shear, sq. mm
	τ	0.91332973	MPa	Shear stress
	OKS	0.91332973		If this space is filled then Double Shear is not a problem
	NOTOKS			If this space is filled then Double Shear is a problem
				FAILURE BY OUT-OF-PLANE INSTABILITY (DISHING)
	K	2		Effective length factor
	π	3.1416		
	r	106.8098	mm	Radius of gyration through thickness direction
	Cc	122.752470		Elastic stability criterion
200000	E		MPa	Modulus of elasticity of lug material
	RATIO	4.58759403		Slenderness ratio
	L	245		
	OK1	4.58759403		If this space is filled then lug will not fail inelastically
	NOTOK1			If this space is filled then lug will fail inelastically
	u	.3		Poisson's Ratio
	Fcrie	−0.00073096	MPa	Inelastic critical buckling stress
	Fcre	103067.187	MPa	Elastic plate buckling stress
	Weff	551	mm	The effective width of lug
	Pd1	−149.02173		Strength of pin-connected plate in inelastic condition, N
	Pd2	2.10123E10		Strength of pin-connected plate in elastic condition, N
	DISHRATIO	0.670817877		Value of DISHRATIO
	OK2	0.670817877		If this space is filled then dishing will not occur
	NOTOK2			If this space is filled the dishing is probable
				BEARING FAILURE
	σB	88.323618	MPa	Bearing failure stress
0	tDP		mm	Thickness of both doubler plates
	OK3	88.323618		If this space is filled then value of bearing stress is acceptable
	NOTOK3			If this space is filled then value of bearing stress is NOT acceptable

FIGURE 7.11(b)   (continued)

   This example shows that the flange top lifting lug is acceptable. It worked in practice on 10 vessels, so the design passed the "acid test."

MULTIPLE LOADS ON LIFTING AND TAIL LUGS

We have now established four modes of failure in lifting lugs as validated in laboratory tests. We now will consider the lugs, and later trunnions, on pressure vessels and stacks exposed to various loads. Consider the schematic in Figure 7.12. Refer to Eq. 1.2 and Eq. 1.3, where the unit of mass, Kg, is converted to the unit of force, N, by Newton's second law.

As the vessel rotates in space in a single plane, the forces L_V, L_H, L_L, T_V, T_H, and T_L vary with the lift angle θ. (In the discussion that follows, any variable name followed by the Greek letter theta, θ, varies with this variable.) Thus, to designate these variables as a function of the lift angle θ, you write them as follows:

$$LV\theta, LH\theta, LL\theta, TV\theta, TH\theta, \text{and } TL\theta$$

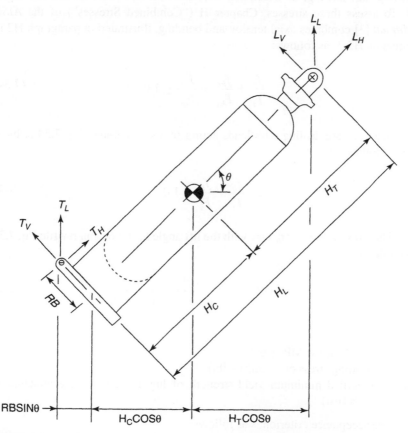

FIGURE 7.12 Lifting schematic of vessel showing forces acting on the vessel. The top lifting lug is a top flange lug.

Using these equations, you make the conversion from mass to force as follows for the top flange lug:

$$P_T = LV\theta\left(\frac{g}{g_c}\right) \text{ Newtons (lb}_f) \tag{7.31}$$

$$P_N = LL\theta\left(\frac{g}{g_c}\right) \text{ Newtons (lb}_f) \tag{7.32}$$

$$P_L = LH\theta\left(\frac{g}{g_c}\right) \text{ Newtons (lb}_f) \tag{7.33}$$

When a stack or vessel is lifted, various loads are imposed on the lift devices (lifting lug, tail lug, or trunnions) over the various values of the lift angle, θ. To assess these stresses, Chapter H ("Combined Stresses") of the *AISC Manual* [3] combines axial tension and bending, illustrated in paragraph H2 in Equation H2-1, as follows:

$$\frac{f_a}{F_t} + \frac{f_{bx}}{F_{bx}} + \frac{f_{by}}{F_{by}} \leq 1.0 \tag{7.34}$$

Since we are dealing with loads acting in a single plane, Eq. 7.34 reduces to the following:

$$\frac{f_a}{F_t} + \frac{f_b}{F_b} \leq 1.0 \tag{7.35}$$

The stress values f_a, f_b vary with the lift angle θ, so you can rewrite Eq. 7.35 as follows:

$$\text{AISCRAT}\theta = \frac{\sigma t\theta}{0.6\sigma_y L} + \frac{\sigma b\theta}{0.66\sigma_y L} \tag{7.36}$$

where
$\sigma t\theta$ = tensile stress MPa (psi)
$\sigma b\theta$ = bending stress N – mm(ft – lb$_f$)
$\sigma_y L$ = specified minimum yield strength of lug and cover plate attachment, MPa (psi)

The acceptance criterion is as follows:

$$\text{AISCRAT}\theta \leq 1 \tag{7.37}$$

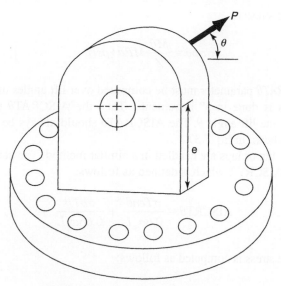

FIGURE 7.13 Lift force acting at an angle θ.

Figure 7.13 illustrates the lift force acting on the lug. From Figure 7.12 and Figure 7.13, you can compute the tensile stress from Eq. 7.33 as follows:

$$P_T = P \sin \theta = L_H \left(\frac{g}{g_c} \right) \tag{7.38}$$

Referring to Figure 7.10, the net tensile area, A_a, is

$$A_a = (W_L)t - (D_H)t = 2b_e t \tag{7.39}$$

where W_L = width of lug = $2b_e + D_H$ mm (in)
    The tensile stress is as follows:

$$\sigma t\theta = \frac{P_T}{A_a} \tag{7.40}$$

The bending stress is computed referring to Eq. 7.31 as

$$M\theta = P(e) \cos \theta = L_H \left(\frac{g}{gc} \right)(e) \tag{7.41}$$

The section modulus of the lifting lug is

$$Zlug = \frac{t_L(W_L^2)}{6} \text{ mm}^3 \text{ (in}^3) \tag{7.42}$$

So the bending stress is

$$\sigma b\theta = \frac{M\theta}{Z} \; MPa\,(psi) \tag{7.43}$$

The *AISCRATθ* parameter must be computed over lift angles of $\theta = 0–90°$. Normally, this is done in 5° increments. Then the AISCRATθ parameter is plotted versus the lift angle θ. The AISCRATθ should always be less than or equal to 1, as shown in Eq. 7.37.

The tailing lug or lugs are handled in a similar method. The parameter with the tail lug is AISCRTθ, which is defined as follows:

$$AISCRT\theta = \frac{\sigma Ten\theta}{0.6\sigma_y L} + \frac{\sigma bT\theta}{0.66\sigma_y L} \tag{7.44}$$

The tensile stress is computed as follows:

$$\sigma Ten\theta = \frac{TV\theta\left(\dfrac{g}{g_c}\right)}{2 Abring} = \frac{TV\theta(9.807)}{2 Abring} \; MPa \tag{7.45}$$

The bending stress is computed as follows:

$$\sigma bT\theta = \frac{MbT\theta}{Ztailug} \; MPa \tag{7.46}$$

The parameters in Eq. 7.45 and Eq. 7.46, *Abring* and *Ztailug*, are computed for the ring block cross-section of the location where the tail lug is welded onto the base plate, compression ring, and skirt area shown in Figure 7.14.

Figure 7.14 shows a spreadsheet solution for the tail lug ring block assembly. The parameter $Abring = Area = 172.6511\ in^2$ and $Ztailug$ = the minimum value of Z1 and Z2, which is $361.88\ in^3$.

The parameter *MbTθ* is defined as follows:

$$MbT\theta = \frac{TV\theta\left(\dfrac{g}{g_c}\right)x + TH\theta\left(\dfrac{g}{g_c}\right)y}{2(1000)} \tag{7.47}$$

In SI metric, this equation becomes

$$MbT\theta = \frac{TV\theta(9.807)x + TH\theta(9.807)y}{2(1000)} \tag{7.48}$$

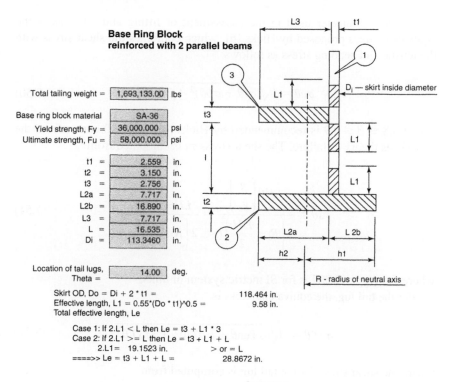

Base Ring Block reinforced with 2 parallel beams

Total tailing weight =	1,693,133.00	lbs

Base ring block material	SA-36	
Yield strength, Fy =	36,000.000	psi
Ultimate strength, Fu =	58,000.000	psi

t1	=	2.559	in.
t2	=	3.150	in.
t3	=	2.756	in.
L2a	=	7.717	in.
L2b	=	16.890	in.
L3	=	7.717	in.
L	=	16.535	in.
Di	=	113.3460	in.

Location of tail lugs, Theta = 14.00 deg.

Skirt OD, Do = Di + 2 * t1 = 118.464 in.
Effective length, L1 = 0.55*(Do * t1)^0.5 = 9.58 in.
Total effective length, Le

Case 1: If 2.L1 < L then Le = t3 + L1 * 3
Case 2: If 2.L1 >= L then Le = t3 + L1 + L
2.L1 = 19.1523 in.   > or = L
====>> Le = t3 + L1 + L = 28.8672 in.

Part #	Width (in.)	Height (in.)	Area (in^2)	Center Line Location (in.)	A * C	A * d^2 (in^4)	I (in^4)
1 Skirt	28.8672	2.5590	73.8710	15.611	1153.16	53.59	93.90
2 Base pl.	3.1500	24.6070	77.5121	12.304	953.67	467.26	4378.42
3 Top rg.	2.7560	7.7170	21.2681	20.749	441.28	763.04	868.58
Sum (Σ) ====>>			172.6511		2548.11		5340.91

h1 = Σ(A*C)/Σ(A) =	14.7587 in.	
h2 =	9.8483 in.	Z1 = Σ(I)/h1 = 361.88 in^3
Radius of neutral axis, R =	57.1007 in.	Z2 = Σ(I)/h2 = 542.32 in^3

FIGURE 7.14 Tail lug ring block assembly.

where
x = distance from tail lug hole center to lug edge, mm
y = radial distance from tail lug hole center to centerline of skirt, mm

The acceptance criterion for AISCRTθ is the same as AISCRATθ, as follows:

$$\text{AISCRT}\theta \leq 1.0 \qquad (7.49)$$

Another parameter used in the assessment of lifting and tail lugs is the equivalent stress proposed by Hicks [6], which combines the shear stress with the tensile and bending stress as follows:

$$\sigma e\theta = \beta \sqrt{(\sigma t\theta + \sigma b\theta)^2 + 3\tau s\theta^2} \qquad (7.50)$$

where $0.8 \le \beta \le 0.9$ is recommended by Hicks [6]. Eq. 7.50 comes from the von Mises theory of failure. The shear stress τs is computed from

$$\tau s\theta = \frac{LH\theta \left(\dfrac{g}{g_c} \right)}{2 \left(R - \dfrac{DH}{2} \right) t} = \frac{LH\theta(9.807)}{2 \left(R - \dfrac{DH}{2} \right) t} \qquad (7.51)$$

where the $g/g_c = 9.807$ is for SI metric system of units.
   For the tail lug, the equivalent stress is

$$\sigma eT\theta = \beta \sqrt{(\sigma \text{Ten}\theta + \sigma bT\theta)^2 + 3\tau sT\theta^2} \qquad (7.52)$$

where the shear stress for the tail lug is computed from

$$\tau sT\theta = \frac{TV\theta \left(\dfrac{g}{g_c} \right)}{4ALUG} = \frac{TV\theta(9.807)}{2ALUG} \qquad (7.53)$$

where, for the tail lug,

$$ALUG = \left(Wtailug - DHtail \right) tlug \qquad (7.54)$$

$Wtailug$ = width of tail lug, mm
$DHtail$ = diameter of hole in tail lug, mm
   $tlug$ = thickness of tail lug, mm

The acceptance criteria for the equivalent stress at the top lug and tail lugs are

$$RATIO\sigma e\theta = \frac{\sigma e\theta}{\sigma_y L} \le 1.0 \qquad (7.55)$$

$$RATIO\sigma eT\theta = \frac{\sigma eT\theta}{\sigma_y L} \le 1.0 \qquad (7.56)$$

Example 7.2 Rigging Analysis of Lifting A Pressure Vessel

This example shows how the lifting and tail lug loadings are evaluated. The vessel being erected is shown in Figure 7.12. We want to compute the reaction loads on the top flange lug and two tail lugs on the bottom. As the vessel is lifted in a two-dimensional plane, the parameters that end with the Greek character θ vary with the lift angle. Figure 7.15(a) shows the equations sheet.

Rules

$$RATLL\theta = \frac{HC \cdot \cosd(\theta) + RB \cdot \sind(\theta)}{HL \cdot \cosd(\theta) + RB \cdot \sind(\theta)}$$

$LL\theta = W \cdot RATLL\theta$

$$RATTL\theta = \frac{HT \cdot \cosd(\theta)}{HL \cdot \cosd(\theta) + RB \cdot \sind(\theta)}$$

$TL\theta = W \cdot RATTL\theta$

$LV\theta = LL\theta \cdot \cosd(\theta)$

$TV\theta = TL\theta \cdot \cosd(\theta)$

$LH\theta = LL\theta \cdot \sind(\theta)$

$TH\theta = TL\theta \cdot \sind(\theta)$

$LVmax = MAX('LV\theta)$

$LHmax = MAX('LH\theta)$

$LLmax = MAX('LL\theta)$

$Lmax = MAX(MAX('LV\theta), MAX('LH\theta))$

$\pi = 3.1416$

$PL = LHmax \cdot 9.807$

$P = LLmax \cdot 9.807$

$PT = LVmax \cdot 9.807$

$$W1\theta = \frac{3 \cdot e \cdot LV\theta}{WL^2}$$

$$W2\theta = \frac{LH\theta}{WL}$$

$P\theta = (W1\theta + W2\theta) \cdot 9.807 \cdot WL$

$M1 = B \cdot PT$

$$Ar\theta = \frac{P\theta}{0.4 \cdot \sigma yL}$$

$Aa = (WL \cdot t) - (DH \cdot t)$

$P\theta max = MAX('P\theta)$

$$Armax = \frac{P\theta max}{0.4 \cdot \sigma yL}$$

If Aa > Armax then Aa = Okay1

If Aa = Armax then Aa = Okay1

If Aa < Armax then Aa = NOTOkay1

;COMPUTING THE AISCRATθ MAXIMUM VALUE FOR THE TOP LIFTING LUG

$$\sigma t\theta = \frac{LH\theta \cdot 9.807}{Aa}$$

$M\theta = LV\theta \cdot 9.807 \cdot e$

FIGURE 7.15(a) The equations sheet for the lift assessment of the vessel in Figure 7.12 This figure is where the spreadsheet lists all the equations used in the algorithm.

$$Zlug = \frac{t \cdot WL^2}{6}$$

$$\sigma b\theta = \frac{M\theta}{Zlug}$$

$$AISCRAT\theta = \frac{\sigma t\theta}{0.6 \cdot \sigma yL} + \frac{\sigma b\theta}{0.66 \cdot \sigma yL}$$

If AISCRATθ < 1.0 then AISCRATθ = Okay2
If AISCRATθ = 1.0 then AISCRATθ = Okay2
If AISCRATθ > 1.0 then AISCRATθ = NOTOkay2

;COMPUTING THE AISCRTθ MAXIMUM VALUE FOR THE TAILING LUG

$$\sigma Ten\theta = \frac{TV\theta \cdot 9.807}{2 \cdot Abring}$$

$$MbT\theta = \frac{TV\theta \cdot 9.807 \cdot x + TH\theta \cdot 9.807 \cdot y}{2 \cdot 1000}$$

$$\sigma bT\theta = \frac{\dfrac{MbT\theta}{Ztailug}}{2 \cdot 1000000}$$

$$AISCRT\theta = \frac{\sigma bT\theta}{0.66 \cdot \sigma yL} + \frac{\sigma Ten\theta}{0.6 \cdot \sigma yL}$$

; COMPUTING EQUIVALENT STRESS FOR THE TOP FLANGE LUG

$$\tau s\theta = \frac{LH\theta \cdot 9.807}{2 \cdot \left[R - \dfrac{DH}{2} \right] \cdot t}$$

$$\sigma e\theta = \beta \cdot \sqrt{(\sigma t\theta + \sigma b\theta)^2 + 3 \cdot \tau s\theta^2}$$

$$RATIO\sigma e\theta = \frac{\sigma e\theta}{\sigma yL}$$

If RATIOσeθ < 1.0 then RATIOσeθ = OKAY3
If RATIOσeθ = 1.0 then RATIOσeθ = OKAY3
If RATIOσeθ > 1.0 then RATIOσeθ = NOTOK3

;COMPUTING EQUIVALENT STRESS IN TAIL LUGS

$$ALUG = (Wtailug - DHtail) \cdot tlug$$

$$\tau sT\theta = \frac{TV\theta \cdot 9.807}{2 \cdot ALUG}$$

$$\sigma eT\theta = \beta \cdot \sqrt{(\sigma Ten\theta + \sigma bT\theta)^2 + 3 \cdot \tau sT\theta^2}$$

$$RATIO\sigma eT\theta = \frac{\sigma eT\theta}{\sigma yL}$$

If RATIOσeTθ < 1.0 then RATIOσeθ = OKAY4
If RATIOσeTθ = 1.0 then RATIOσeθ = OKAY4
If RATIOσeTθ > 1.0 then RATIOσeθ = NOTOK4

FIGURE 7.15(a)   (continued)

Status	Input	Name	Output	Unit	Comment
L		RATLLθ	.489801255		
	18730	HC		mm	Length from tailling lug to Center of Gravity (CG)
L	0	θ			Lift angle in degrees
	3309.5	RB		mm	Outside radius of skirt center line to tail lug hole center
	38240	HL		mm	Length from tailling lug to top lifting lug (HC + HT)
L		LLθ	647601.726	Kg	
	1322172.45	W		Kg	Lifting weight - includes dynamic load factor
L		RATTLθ	.510198745		
	19510	HT		mm	Length from top lifting lug to CG
L		TLθ	674570.724	Kg	Vertical lift load component at each tail lug varies with θ
L		LVθ	647601.726	Kg	Tangential lift load component at top lift lug- varies with θ
L		TVθ	674570.724	Kg	Tangential lift load component at each tail lug - varies with θ
L		LHθ	0	Kg	
L		THθ	0	Kg	
		LVmax	650187.413	Kg	Max value of LVθ
		LHmax	1322172.45	Kg	Max value of LHθ
		LLmax	1322172.45	Kg	Max value of LLθ
		Lmax	1322172.45	Kg	Max load value
		π	3.1416		
		PL	12966545.2	N	
		P	12966545.2	N	
		PT	6376387.96	N	
L		W1θ	1343.4814		
	410	e			Vertical distance from flange surface to centerline of lug hole
	770	WL		mm	Width of top lug plate
L		W2θ	0		
L		Pθ	10145152		
		M1	4846054849		
	760	B		mm	Vert. Dist. from cover plate bottom to lug hole centerline
L		Arθ	102269.678		Required tensile area of top lug for each value of θ
	248	σyL		MPa	Specified minimum yield strength of lug
		Aa	163760		Calculated actual tensile area, sq mm
	356	t		mm	Thickness of top lug plate

FIGURE 7.15(b) The variables sheet showing all the variables used for the lift assessment. This figure is where the spreadsheet lists all the variables used in the algorithm.

Status	Input	Name	Output	Unit	Comment
	310	DH		mm	Diameter of hole in top lug plate
		Armax	130711.141		Maximum required tensile area of top lifting lug
		Okay1	163760		If this space is filled then the bolt area is satisfactory
		NOTOkay1			If this space is filled then the bolt area is NOT satisfactory
					COMPUTING THE AISCRATθ MAX VALUE
		Pθmax	12966545.2	N	
L		σtθ	0	MPa	Tensile stress in top lifting lug - varies with lift angle θ
L		Mθ	2603922351		Bending moment in top lifting lug, N-mm
		Zlug	35178733.3		Section modulus of cross section of top lifting lug, mm^3
L		σbθ	74.0197871	MPa	Bending stress in top lifting lug - varies with lift angle θ
L		AISCRATθ	.45222255		AISC Ratio for top lifting lug - see plot for values
		Okay2	.45222255		If this space is filled then the AISCRATθ max value is OK
		NOTOkay2			If this space is filled then the AISCRATθ max value is NOT OK
					COMPUTING THE MAXIMUM AISCRTθ VALUE
		σTenθ	29.695808	MPa	Tensile stress in tail lug - varies with lift angle θ
	111388.03	Abring			Area of ring block (Fig. 7.12), sq mm
		MbTθ	694629.085	MPa	Bending stress in the tail lugs - varies with θ, N-mm
	210	x		mm	Distance from tail lug hole to lug edge
	417.5	y		mm	Radial distance from tail lug hole center to centerline of skirt
	.006	Ztailug			Section modulus of ring block (Fig. 7.12), cu meters
L		σbTθ	57.8857571		Bending stress in tail lugs - varies with lift angle θ
L		AISCRTθ	.553220588		AISC Ratio for tail lugs - see plot for values
					COMPUTING EQUIVALENT STRESS IN TOP FLANGE LUG
L		τSθ	0	MPa	Shear stress in top flange lug
	385	R		mm	Radius of top lug extending from lug hole center to lug edge

FIGURE 7.15(b)   (continued)

Status	Input	Name	Output	Unit	Comment
L		σeθ	66.6178084	MPa	Equivalent stress in top flange lug
	.9	β			Factor used in computation of equivalent stress
L		RATIOσeθ	.268620195		
		OKAY3	.268620195		If this space is filled then RATIOσeθ is satisfactory
		NOTOK3			If this space is filled then RATIOσeθ is NOT satisfactory
					COMPUTING EQUIVALENT STRESS IN TAIL LUGS
		ALUG	39750		Shear area of each of two tail lugs, sq mm
	420	Wtailug		mm	Width of each of two tail lugs
	155	DHtail		mm	Diameter of hole in each tail lug
	150	tlug		mm	Thickness of each tail lug
L		TSTθ	83.2140263	MPa	Shear stress in each tail lug
L		σeTθ	151.788817	MPa	Equivalent stress in each tail lug
L		RATIOσeTθ	.612051682		
		OKAY4	.268620195		If this space is filled then RATIOσeTθ is acceptable
		NOTOK4			If this space is filled then RATIOσeTθ is NOT acceptable

FIGURE 7.15(b) (continued)

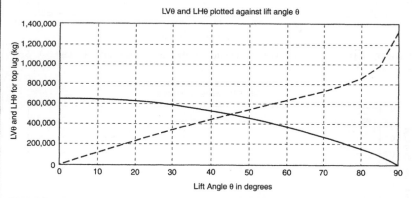

FIGURE 7.16 The loads LVθ (solid line) and LHθ (dashed line) versus θ for top flange lug. Refer to Figure 7.12 for the orientation of these forces.

Figures 7.16 through 7.21 are plots of the various ratios used in the algorithm. The reader can tell at a glance if the ratios are acceptable. The corresponding equation number is indicated below each figure.

The actual lifting of the hydrocracker in this example is shown in Figures 7.22 and 7.23. This is validation that the analysis works, as there were 10 similar vessels lifted and installed on site.

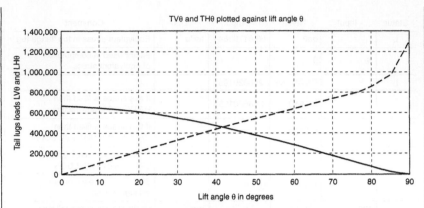

FIGURE 7.17 The loads TVθ (solid line) and THθ (dashed line) versus θ for top flange lug. Refer to Figure 7.12 for the orientation of these forces.

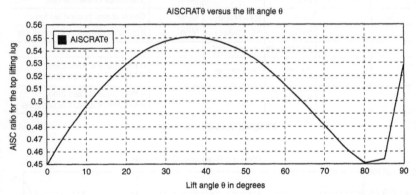

FIGURE 7.18 The AISCRATθ for the top flange lug versus the lift angle θ. See Eq. 7.37.

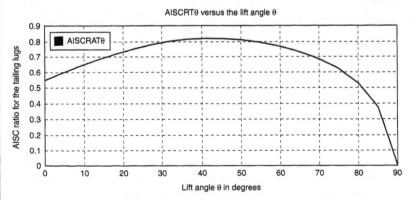

FIGURE 7.19 The AISCRTθ for each tail lug versus the lift angle θ. See Eq. 7.44.

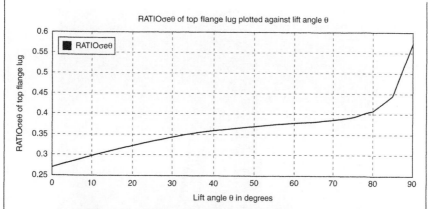

FIGURE 7.20   RATIO$\sigma e\theta$ of top flange lug versus θ. See Eq. 7.53.

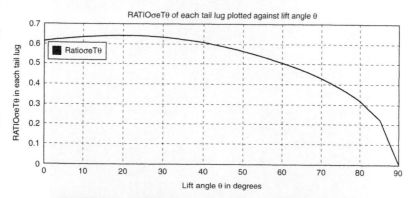

FIGURE 7.21   RATIO$\sigma eT\theta$ for each tail lug versus θ. See Eq. 7.54.

FIGURE 7.22   The actual lifting of the hydrocracker using the top flange lug and tail lugs.

FIGURE 7.23 The lowering and final installation of the hydrocracker referred to in Figure 7.22.

TRUNNIONS

Trunnions are used for vessels that are too tall to be lifted by lugs at the top head. Figure 7.24 shows the three basic types of trunnions.

Unlike lifting lugs, it is not as common for trunnions to be designed and fabricated by an operating facility. Trunnions are mostly designed by an engineering contractor or a fabrication shop and welded into place in the fabrication shop. Trunnions are used to erect vessels that are too tall to be lifted by lugs. Lifting from lugs at top would result in excessive bending stresses and possible distortion in the shell. That is why we will not go into much detail here; this book is intended for field applications, although much of the discussion and examples are equally applicable to engineering companies and fabricators. However, it is important for field personnel to understand trunnions and how to assess them—contractors do make mistakes.

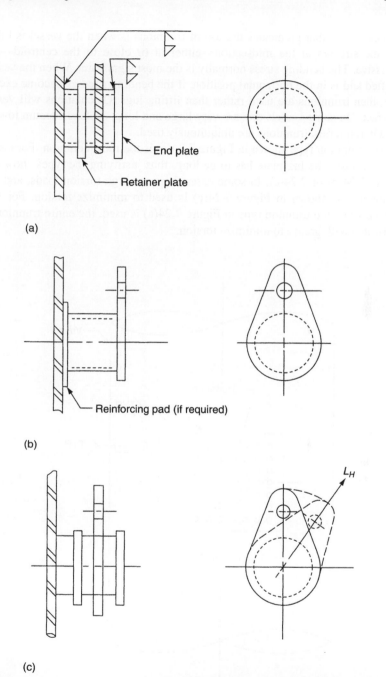

FIGURE 7.24 The three basic types of trunnions: **(a)** trunnion only; **(b)** trunnion and fixed lug; **(c)** trunnion and rotating lug.

One factor that predicates the use of trunnions is when the vessel is lifted and the stresses at the midsection—either at or close to the centroid—get excessive. The bending stress normally is the most significant. When the vessel is lifted and is in the horizontal position, if the bending stresses become excessive, then trunnions are used rather than lifting lugs. Calculations will verify this fact. On some vessels that are very large and long, such as vacuum towers used in refineries, trunnions are ubiquitously used.

The trunnion type shown in Figure 7.24(a) is the most common. For clearance purposes the lug plate has to be long, thus justifying the types shown in Figure 7.24(b) or 7.24(c). In some cases there are high torsion loads, and the trunnion type shown in Figure 7.24(c) is used to minimize torsion. For this reason, when the trunnion type in Figure 7.24(a) is used, the entire trunnion is lubricated with grease to minimize torsion.

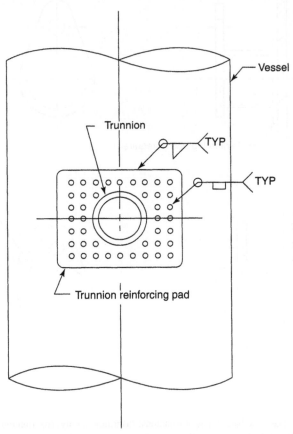

FIGURE 7.25 A trunnion reinforcing pad with plug welds. Along with the weld connecting the reinforcing pad to the shell, the welds in the plug hole add considerably to the weld strength. The trunnion is welded both to the vessel bare wall and the reinforcing pad.

There are three cardinal rules about trunnions, which are as follows:

1. The trunnion cylinder should be welded to the vessel shell and the reinforcing pad.
2. If the trunnion load is sufficiently high to warrant a reinforcing pad, and a conventional circular pad is not adequate, then an enlarged pad should be used with plug hole welds. The plug hole welds will multiply the weld area.
3. The trunnion support should be analyzed like lifting lugs with the lift angle, θ, varying from 0–90°.

If a trunnion is welded to a pad, which in turn is welded to the vessel shell with fillet welds, the fillet welds take the entire loads and consequently may shear off the vessel during lift. An enlarged reinforcing pad with plug welds will multiply the weld area, while at the same time the trunnion is welded to the vessel base metal. This type of arrangement is shown in Figure 7.25.

Another device used with high-lifting loads using a trunnion is a gusset plate, as shown in Figure 7.26.

FIGURE 7.26 A trunnion with gusset plates welded to the reinforcing pad. The reinforcing pad cannot be seen because of the insulation and aluminum jacket covering the vessel. The vessel operates at high temperatures because it is an FCCU regenerator.

Example 7.3 Erection Analysis Using Trunnions

This example illustrates the erection analysis of a vacuum tower in a refinery that has trunnions. The lifting schematic is shown in Figure 7.27.

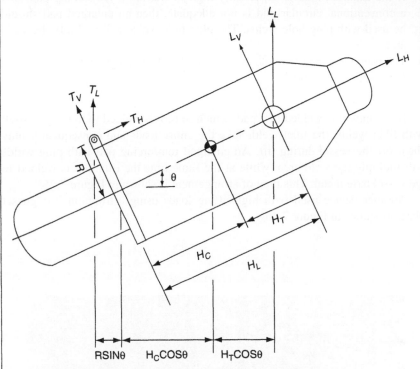

FIGURE 7.27   The lifting schematic of a vacuum tower with trunnions.

$$RATLL\theta = \frac{HC \cdot cosd(\theta) + R \cdot sind(\theta)}{HL \cdot cosd(\theta) + R \cdot sind(\theta)}$$

$LL\theta = W \quad RATLL\theta$

$$RATTL\theta = \frac{HT \cdot cosd(\theta)}{HL \cdot cosd(\theta) + R \cdot sind(\theta)}$$

$TL\theta = W \cdot RATTL\theta$

$LV\theta = LL\theta \cdot cosd(\theta)$

$TV\theta = TL\theta \cdot cosd(\theta)$

$LH\theta = LL\theta \cdot sind(\theta)$

$TH\theta = TL\theta \cdot sind(\theta)$

$LVmax = MAX ('LV\theta)$

$LHmax = MAX ('LH\theta)$

$LLmax = MAX ('LL\theta)$

$Lmax = MAX (MAX('LV\theta), MAX('LH\theta))$

$$F = \frac{Lmax \cdot 9.807}{N}$$

$\pi = 3.1416$

$$Z = \frac{\left[\dfrac{\pi}{32}\right] \cdot [do^4 - di^4]}{do}$$

$$FH\theta = \frac{LH\theta \cdot 9.807}{N}$$

$ftball = 0.66 \cdot SMYS$

$$ftbL\theta = \frac{FH\theta \cdot e}{Z}$$

$$FV\theta = \frac{LV\theta \cdot 9.807}{N}$$

$$ftbC\theta = \frac{FV\theta \cdot e}{Z}$$

$$TAREA = \left[\frac{\pi}{4}\right] \cdot \left[do^2 - di^2\right]$$

$$ftCP\theta = \frac{FV\theta}{TAREA}$$

$ftCPall = 0.6 \cdot SMYS$

$$AISCT\theta = \frac{ftbL\theta + ftbC\theta}{ftball} + \left[\frac{ftCP\theta}{ftCPall}\right]$$

If $AISCT\theta < 1.0$ then $AISCT\theta = OK1$

If $AISCT\theta = 1.0$ then $AISCT\theta = OK1$

If $AISCT\theta > 1.0$ then $AISCT\theta = NOTOKA$

$$fsT\theta = \frac{LH\theta}{TAREA}$$

$fsall = 0.4 \cdot SMYS$

$fsTmax = MAX('fsT\theta)$

FIGURE 7.28 The equations sheet for the lifting analysis of the vacuum. This figure is where the spreadsheet lists all the equations used in the algorithm.

If fsTθ < fsall then fsTθ = OK2

If fs Tθ = fsall then fsTθ = OK2

If fs Tθ > fsall then fsTθ = NOTOK2

$$\sigma eT\theta = \sqrt{(ftbL\theta + ftbC\theta + ftCP\theta)^2 + 3 \cdot fsT\theta^2}$$

$$RAT\sigma eT\theta = \frac{\sigma eT\theta}{SMYS}$$

$$Mbtail\theta = \frac{TV\theta \cdot 9.807 \cdot x + TH\theta \cdot 9.807 \cdot y}{2 \cdot 1000}$$

$$\sigma btail\theta = \frac{\dfrac{Mbtail\theta}{Ztailug}}{1000000}$$

$$\sigma Tent\theta = \frac{TV\theta \cdot 9.807}{2 \cdot Abring}$$

$$ALUG = WLt \cdot ttail - Dhole \cdot ttail$$

$$\tau tail\theta = \frac{TV\theta \cdot 9.807}{4 \cdot ALUG}$$

$$\sigma eTail = \sqrt{(\sigma Tent\theta + \sigma btail\theta)^2 + 3 \cdot \tau tail\theta^3}$$

$$AISCRT\theta = \left[\frac{\sigma Tent\theta}{0.6 \cdot \sigma yL}\right] + \frac{\sigma btail\theta}{0.66 \cdot \sigma yL}$$

$$RAT\sigma eT = \frac{\sigma eTail}{\sigma yL}$$

$$RtLug = \frac{WLt}{2}$$

$$HA = \left[RtLug - \frac{Dhole}{2}\right] \cdot ttail$$

$$TVt = MAX('TV\theta)$$

$$PVt = TVt \cdot 9807$$

$$\sigma HT = \frac{PVt}{Nt \cdot HA}$$

If σHT < 0.45 · σyL then σHT = OK

If σHT = 0.45 · σyL then σHT = OK

If σHT = 0.45 · σyL then σHT = NOTOK

;COMPUTING STRESS IN TRUNNION WELDS

$$\sigma bw\theta = \frac{5.66 \cdot LL\theta \cdot e}{h \cdot do^2 \cdot \pi} + \frac{5.66 \cdot LL\theta \cdot e}{hp \cdot Dp^2 \cdot \pi}$$

$$\tau bw\theta = \frac{2.83 \cdot MT}{\pi \cdot h \cdot do^2} + \frac{2.83 \cdot MT}{\pi \cdot hp \cdot Dp^2}$$

$$\sigma w\theta = 0.9. \sqrt{\sigma bw\theta^2 + 3 \cdot Tbw\theta^2}$$

$$AISCW\theta = \frac{\sigma w\theta}{\sigma yL}$$

FIGURE 7.28   (continued)

Variables Sheet

Status	Input	Name	Output	Unit	Comment
L		RATLLθ	0.31938369		
	9639	HC		mm	Length from tailing lug to Center of Gravity (CG)
L	0	θ		deg	Lift angle
	6755	R		mm	Outside radius of center line of skirt to center of tail lug hole
	30180	HL		mm	Length from tailing lug to trunnion (HC + HT)
L		LLθ	268282.306	Kg	Vertical lift force component at trunnion - varies with θ
	840000	W		Kg	Lift weight includes dynamic load factor
L		RATTLθ	0.680616302		
	20541	HT		mm	Length from trunnion to CG
L		TLθ	571717.694	Kg	Vertical lift force component at tail lug - varies with θ
L		LVθ	268282.306	Kg	Lift weight component perpendicular to vert axis at trunnion
L		TVθ	571717.694	Kg	Lift force component at tail lug perpendicular to vert axis
L		LHθ	0	Kg	Top lift force component coincident with vessel vert axis
L		THθ	0	Kg	Tail lift weight component coincident with vert axis
		LVmax	292572.316	Kg	Maximum weight component at top perpendicular to vert axis at 20 degrees
		LHmax	840000	Kg	Maximum weight component at top parallel to vessel axis at 90 degrees
		LLmax	840000	Kg	Maximum vertical weight component at top at 90 degrees
		Lmax	840000	Kg	Maximum weight on lifting trunnion
					ANALYSIS FOR TRUNNIONS
	2	N			Number of trunnions
		F	4118940	N	Maximum force on one lifting trunnion
		π	3.1416		
		Z	42018490.9		Section modulus of trunnion cross section, cu mm
	1500	do		mm	Outside diameter of trunnion
	1450	di		mm	Inside diameter of trunnion
		ftb		MPa	Longitudinal Bending stress in trunnion
	575	e		mm	Length of trunnion
		ftball	171.6		Allowable bending stress of trunnion
	260	SMYS			Specified minimum yield stress of trunnion material (SA-516-70)
		TAREA	115846.5		Trunnion cross sectional area, sq mm
L		ftbLθ	0	MPa	Maximum longitudinal bending stress on trunnion
L		ftbCθ	18.002200	MPa	Maximum circumferential bending stress on trunnion
L		FHθ	0	N	Force component due to LH

FIGURE 7.29 The variable sheet for the vacuum tower lifting analysis. The variable sheet is where the spreadsheet lists all the variables used in the algorithm. The computations of *Abring* and *Ztailug* are shown in Figure 7.30.

Status	Input	Name	Output	Unit	Comment
L		FVθ	1315522.29	N	Force component due to LV
		ftCPθ	11.355736	MPa	Compressive stress on trunnion
		ftCPall	156	MPa	Allowable compressive stress
		AISCTθ	0.177701111		AISC Allowable Ratio for bending and shear per Section 5, Chapter H
		OK1	0.177701111		
		NOTOKA			
L		fsTθ	0	MPa	Shear stress at trunnion vessel wall
		fsall	104	MPa	Allowable shear stress
		fsTmax	7.25097435	MPa	Maximum shear stress at trunnion & vessel wall
		OK2	0		If this space contains an entry then shear stress is acceptable
		NOTOK2			If this space contains an entry then shear stress is NOT acceptable
L		σeTθ	29.357937	MPa	Von Mises equivalent stress for trunnion
L		RATσeTθ	0.112915143	MPa	
					ANALYSIS FOR TAIL LUGS
	350	x		mm	Distance from tail lug hole center to lug edge
	484	y		mm	Radial distance from tail lug hole center to centerline of skirt
L		Mbtailθ	981196.199	Nm	Bending moment at tail lug attachment
L		σbtailθ	87.606803	MPa	Bending stress in each tail lug
	.0112	Ztailug			Section modulus of ring assembly from Excel spreadsheet, cu m
	98876	Abring			Tensile stress area of each tail lug, sq mm
		ALUG	61200		Shear area of each tail lug, sq mm
	700	WLt		mm	Width of each tail lug
	120	ttail		mm	Thickness of each tail lug
	190	Dhole		mm	Diameter of hole in each tail lug
L		τtailθ	22.903739	MPa	Shear stress in each tail lug
L		σeTail	222.466320	MPa	Von Mises equivalent stress in each tail lug
L		σTentθ	28.352863	MPa	Tensile stress in each tail lug
L		AISCRTθ	0.692278282		AISC Ratio for each tail lug based on Sec. 5 Chapter H
	260	σyL			Yield stress of lug plate (SA-516-70)
L		RATσeT	0.855639691		Ratio of equivalent stress of SMYS
					CHECKING HOOP TENSILE STRESS IN TAIL LUGS
		RtLug	350	mm	Radius of tail lug from center to edge of lug
		HA	30600		Hoop tensile area, sq mm

FIGURE 7.29   (continued)

Status	Input	Name	Output	Unit	Comment
		TVt	571717.694		Maximum tensile weight on both tail lugs
		PVt	5606835.423459		Maximum tensile load on both tail lugs
		σHT	91.6149579	MPa	Hoop Tensile stress on each tail lug
	2	Nt			Number of tail lugs
		OK	91.6149579		If this space is filled than Hoop Tension stress is OK
		NOTOK			If this space is filled than Hoop Tension stress is NOTOK
					COMPUTING WELD STRESSES ON TRUNNION
L		σbwθ	24.125312	MPa	Bending stress in trunnion-pad configuration
	8	h		mm	Size of weld connecting trunnion to pad or shell
	8	hp		mm	Size of weld connecting trunnion pad to shell
	2000	Dp		mm	Diameter of trunnion pad
L		τbwθ	0	MPa	Shear stress in trunnion-pad configuration
	0	MT			Torsion moment on trunnion = 0 because of grease
		σwθ	21.7127808		
L		AISCWθ	0.083510696		

FIGURE 7.29 (continued)

Base Ring Block

reinforced with 2 parallel beams

Total Tailing Weight = [571,717.70] kg

Base Ring Block Material [5A-36]
Yield Strength, Fy = [248.000] MPa
Ultimate Strength, Fu = [399.896] MPa

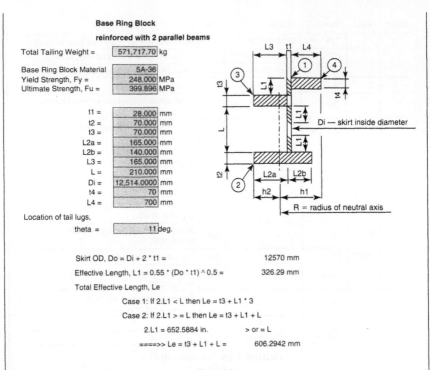

t1 = [28.000] mm
t2 = [70.000] mm
t3 = [70.000] mm
L2a = [165.000] mm
L2b = [140.000] mm
L3 = [165.000] mm
L = [210.000] mm
Di = [12,514.0000] mm
t4 = [70] mm
L4 = [700] mm

Location of tail lugs,

theta = [11] deg.

Skirt OD, Do = Di + 2 * t1 = 12570 mm

Effective Length, L1 = 0.55 * (Do * t1) ^ 0.5 = 326.29 mm

Total Effective Length, Le

Case 1: If 2.L1 < L then Le = t3 + L1 * 3

Case 2: If 2.L1 > = L then Le = t3 + L1 + L

2.L1 = 652.5884 in. > or = L

====>> Le = t3 + L1 + L = 606.2942 mm

Center Line

Part #	Width (mm)	Height (mm)	Area (mm^2)	Location (mm)	A * C	A * d^2 (mm^4)	I (mm^4)
1 skirt	606	28	16976	714	12121033	453217268	454326382
2 base pl.	70	305	21350	741	15809675	769864279	935371258
3 top rg.	70	165	11550	811	9361275	780135362	806339424
4 Top stiff	70	700	49000	350	17150000	1971922224	3972755557
Sum (Σ) ====>>			98876		54441983		6168792622

h1 = Σ(A*C)/Σ(A) = 550.6074 mm   Z1 = Σ(I)/h1 = **11203615** mm^3
h2 = 342.3926 mm   Z2 = Σ(I)/h2 = **18016721** mm^3
Radius of Neutral Axis, R = **6696** mm

FIGURE 7.30 Ring block section properties for vacuum tower, where Abring = 98876 mm³ and Ztailug = 11203615mm³ = 0.0112m³.

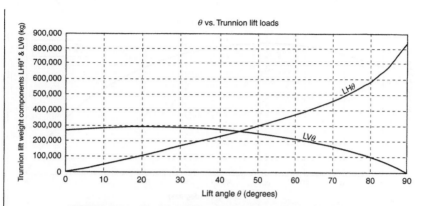

FIGURE 7.31 Trunnion lift loads LVθ and LHθ plotted against θ.

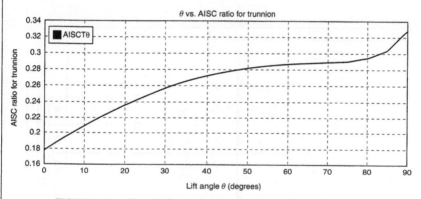

FIGURE 7.32 The AISC ratio for the trunnions versus the lift angle θ.

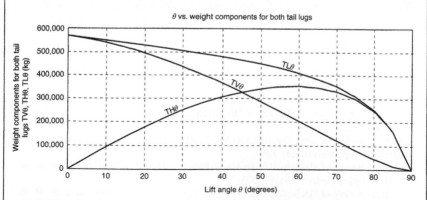

FIGURE 7.33 Weight components TVθ, TLθ, and THθ plotted against the lift angle θ.

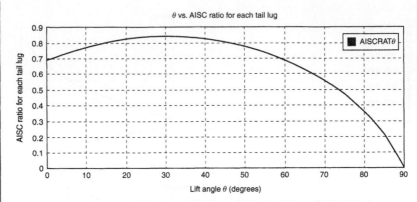

FIGURE 7.34   AISC ratio for each tail lug plotted against the lift angle θ.

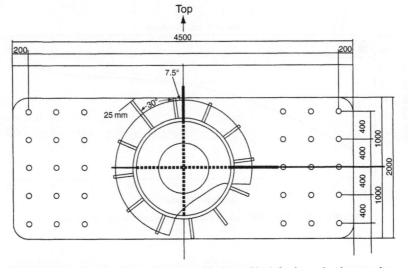

FIGURE 7.35   Drawing of actual trunnion with plug weld reinforcing pad and gusset plates.

REFERENCES

1. Duerr, David, "Pinned Connection Strength and Behavior," *Journal of Structural Engineering ASCE*, February 2006, pp 183–194.
2. Bragassa, P. W., "Tank TSF-09/18 V-Tanks Remediation Tank Lifting Design, Document ID: EDF-5595, Project No. 22901," Idaho National Engineering and Environmental Laboratory, April 18, 2005.
3. American Institute of Steel Construction (AISC), *AISC Manual of Steel Construction*, 9th edition, Chicago, Illinois, July, 1989.
4. American Society of Mechanical Engineers (ASME), *ASME BTH-1-2005 Design of Below-the-Hook Lifting Devices*, ASME, 2005.
5. American Society of Mechanical Engineers (ASME), *ASME B30.20-2006 Below-the-Hook Lifting Devices*, ASME, 2006.
6. Hicks, John, *Weld Joint Design*, 3rd edition, Industrial Press, Inc., New York, 1999.

Chapter 8

Assessing Weld Attachments

The assessment of welds attaching lifting lugs, tail lugs, and trunnions to vessels and stacks is a very important function. A simple lifting lug is shown in Figures 8.1(a) and (b). The lug in Figure 8.1(a) is attached to the vessel or stack with simple "U" bend welding—two fillet welds on each side and one fillet weld at the bottom of the lug. The lug in Figure 8.1(b) is attached to the vessel or stack by two fillet welds on the outside, two fillet welds on the two bottom portions, two fillet welds on the inside of the cutout, and the fillet weld on the arc at the top of the cutout. The pattern shown in Figure 8.1(b) is used when more welding strength is needed.

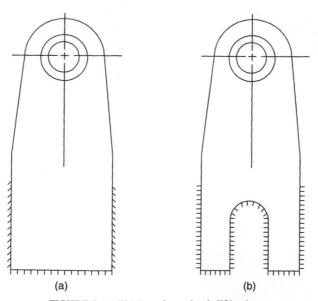

(a)   (b)

FIGURE 8.1   Welding of two simple lifting lugs.

Example 8.1 Evaluation of the Welds in a Simple "U" Configuration

Here, we will analyze the configuration shown in Figure 8.1(a). Refer now to Figure 8.2.

Figure 8.3(a) shows the equations used in assessing these welds, and Figure 8.3(b) shows the variables and solutions. As you can see from Figure 8.3(b), the welds are satisfactory.

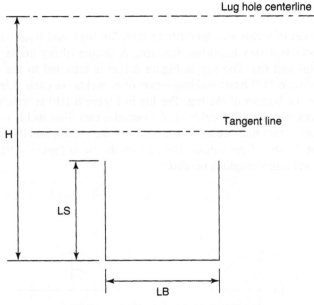

FIGURE 8.2   Lifting lug with simple "U" design.

;CALCULATION OF "U" SHAPED WELD ON LUG

$$P2 = \frac{P1 \cdot IF}{2}$$

$$P3 = \frac{P1 \cdot IF \cdot HC}{2 \cdot HL}$$

$$Aw = (LB + 2 \cdot LS) \cdot 0.707 \cdot h$$

$$Th = \frac{P3}{Aw}$$

$$LCG = \frac{LS^2}{2 \cdot LS + LB}$$

$$Jw = \left[\frac{8 \cdot LS^3 + 6 \cdot LS \cdot LB^2 + LB^3}{12} \right] - \left[\frac{LS^4}{2 \cdot LS + LB} \right]$$

$$J = 0.707 \cdot h \cdot Jw$$

$$Mw = P3 \cdot (H-LCG)$$

$$\tau Wa = \frac{Mw \cdot \left[\frac{LB}{2} \right]}{J}$$

$$\tau WL = \frac{Mw \cdot (LS - LCG)}{J}$$

$$\tau TOT = \tau h + \tau WL$$

;STRESS IN WELD WITH VESSEL IN VERTICAL POSITION

$$\sigma w1 = \frac{P2}{Aw}$$

$\sigma w1all = 0.4 \cdot \sigma yw$

If $\sigma w1 < \sigma w1all$ then $\sigma w1 = $ OKAY1

If $\sigma w1 = \sigma w1all$ then $\sigma w1 = $ OKAY1

If $\sigma w1 > \sigma w1all$ then $\sigma w1 = $ NOTOK1

;STRESS IN WELD WITH VESSEL IN HORIZONTAL POSITION

$$\sigma w2 = \sqrt{\tau Wa^2 + \tau TOT^2}$$

$\sigma w2all = 0.4 \cdot \sigma yw$

If $\sigma w2 < \sigma w2all$ then $\sigma w2 = $ OKAY2

If $\sigma w2 = \sigma w2all$ then $\sigma w2 = $ OKAY2

If $\sigma w2 = \sigma w2all$ then $\sigma w2 = $ NOTOK2

FIGURE 8.3(a)   The equations sheet showing the equations used in the assessment of the welds.

Input	Name	Output	Unit	Comment
	P2	375,000	lbm	Load on lug in vertical position
500,000	P1		lbm	Weight of vessel
1.5	IF			Impact Factor (Normally 1.5)
	P3	215,000	lbm	Load on lug in horizontal position
860	HC		in	Distance from tailing lug hole to CG
1500	HL		in	Distance from hole in tailing lug to hole in top lifting lug
	Aw	120.19		Weld area, sq in
28	LB		in	Length of lifting lug on bottom
20	LS		in	Weld height on sides
2.5	h		in	Lug weld leg size
	τh	1788.834346	psi	Shear stress in weld in horizontal position
	LCG	5.882353	in	Distance to weld centroid
	Jw	12649.725490		Weld unit polar moment of inertia, in^3
	J	22358.389804		Weld polar moment of inertia, in^4
	Mw	6475294.117647	in-lb	Moment in weld
36	H		in	Height from bottom of lug to centerline of lug hole
	τWa	4054.590623	psi	Weld axial torsional stress
	τWL	4088.662813	psi	Weld lateral torsional stress
	τTOT	5877.497159	psi	Total lateral weld shear stress
				STRESS IN WELD WITH VESSEL IN VERTICAL POSITION
	σw1	3120.059905	psi	Stress in weld with vessel in horizontal position
38000	σw1all		psi	Allowable stress in weld for vertical position
	oyw	95,000	psi	Specified minimum yield stress of weld metal
	OKAY1	3120.059905		If this space is filled then vertical weld stress is OK
	NOTOK1			If this space is filled then vertical weld stress is NOTOK
				STRESS IN WELD WITH VESSEL IN HORIZONTAL POSTION
	σw2	7140.355592	psi	Stress in weld with vessel in horizontal postion
38000	σw2all		psi	Allowable stress in weld for vertical position
	OKAY2	7140.355592		If this space is filled then horizontal weld stress is OK
	NOTOK2			If this space is filled then horizontal weld stress is NOTOK

FIGURE 8.3(b) The variable sheet with the variables and solutions.

Example 8.2 Lifting Lug with "U" Shape and Cutout

Figure 8.4 illustrates a simple lifting lug with a cutout for more welding area.

The solution to this kind of lug weld configuration is shown in Figures 8.5(a) and 8.5(b). As you can see in Figures 8.5(a) and 8.5(b), the welds are satisfactory for this application.

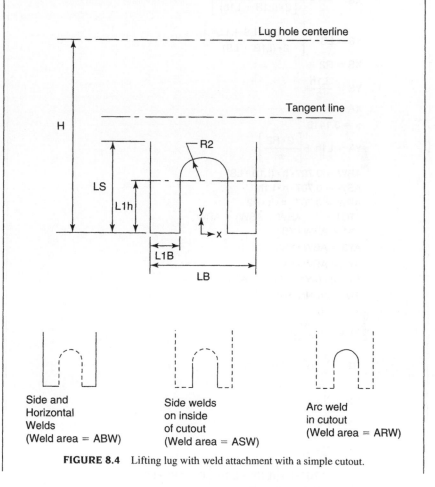

Side and
Horizontal
Welds
(Weld area = ABW)

Side welds
on inside
of cutout
(Weld area = ASW)

Arc weld
in cutout
(Weld area = ARW)

FIGURE 8.4   Lifting lug with weld attachment with a simple cutout.

$$P2 = \frac{P1 \cdot IF}{2}$$

$$P3 = \frac{P1 \cdot IF \cdot HC}{2 \cdot HL}$$

$$L1B = \frac{LB}{2} - R2$$

$$XB = \frac{LB}{2} - \left[\frac{L1B^2}{2 \cdot (L1B + L1h)}\right]$$

$$YB = LS - \left[\frac{2 \cdot L1B \cdot LS + LS^2}{2 \cdot (L1B + LS)}\right]$$

$$XS = R2$$

$$YS = \frac{L1h}{2}$$

$$XA = 0$$

$$\pi = 3.1416$$

$$YA = L1h + \left[\frac{2 \cdot R2}{\pi}\right]$$

$$ABW = 0.707 \cdot h \cdot (L1B + LS)$$

$$ASW = 0.707 \cdot h \cdot L1h$$

$$ARW = 0.707 \cdot \pi \cdot h \cdot R2$$

$$ATOT = 2 \cdot (ABW + ASW) + ARW$$

$$AY1 = ABW \cdot YB$$

$$AY3 = ASW \cdot YS$$

$$AY5 = ARW \cdot YA$$

$$AY = 2 \cdot (AY1 + AY3) + AY5$$

;By symmetry Xbar = 0

$$Xbar = 0$$

$$Ybar = \frac{AY}{ATOT}$$

$$R1 = \sqrt{XB^2 + (Ybar - YB)^2}$$

$$R3 = \sqrt{XS^2 + (Ybar - YS)^2}$$

$$R5 = YA - Ybar$$

$$S3 = \frac{P3}{ATOT}$$

$$JB = \frac{0.707 \cdot h \cdot \left[(L1B + LS)^4 - \left[6 \cdot L1B^2 \cdot LS^2\right]\right]}{12 \cdot (L1B + LS)}$$

$$JS = \frac{0.707 \cdot h \cdot L1h^3}{12}$$

$$JR = 0.707 \cdot h \cdot \pi \cdot R2^3$$

$$J = 2 \cdot \left[JB + JS + ABW \cdot R1^2 + ASW \cdot R3^2\right] + JR + ARW \cdot R5^2$$

FIGURE 8.5(a)   The equations sheet for the weld assessment.

M2 = P3 • (H-Ybar)

$$\tau y = \frac{M2 \cdot \left[\dfrac{LB}{2} \right]}{J}$$

C1 = LS−Ybar

C2 = Ybar

If C1 > C2 then C = C1

If C2 > C1 then C = C2

$$\tau_{xx} = \frac{M2 \cdot C}{J}$$

$\tau_w = S3 + \tau_{xx}$

;SOLVING FOR STRESS IN WELD WITH VESSEL IN VERTICAL POSITION

$$S1 = \frac{P2}{ATOT}$$

σ1 all = 0.4 • σyw

If S1 < σ1all then S1 = Okay1

If S1 = σ1all then S1 = Okay1

If S1 > σ1all then S1 = NOTOK1

;SOLVING FOR STRESS IN WELD WITH VESSEL IN HORIZONTAL POSITION PC

$$S2 = \sqrt{\sigma y^2 + \sigma w^2}$$

If S2 < σ1all then S2 = OKay2

If S2 = σ1all then S2 = OKay2

If S2 > σ1all then S2 = NOTOK2

FIGURE 8.5(a)   (continued)

Input	Name	Output	Unit	Comment
	P2	450,000	lb	Load on lug with vessel in vertical position
600,000	P1		lb	Lift weight
1.5	IF			Impact Factor (1.5 is recommended)
	P3	263268.443	lb	Load on lug with vessel in horizontal position
856.5	HC		in	Distance from hole in tail lug to vessel CG
1464	HL		in	Distance from hole in tail lug to hole in lift lug
	L1B	10	in	Bottom weld leg length
24	LB		in	Lug width at bottom
2	R2		in	Radius of arc weld made on lug in cutout
	XB	9.5	in	Centroid of side and horizontal bottom welds in x direction
10	L1h		in	Weld cutout height
	YB	5.78571429	in	Centroid of side and horizontal bottom welds in y direction
18	LS		in	Side weld height
	XS	2	in	Centroid of side and horizontal bottom welds in x direction
	YS	5	in	Centroid of side and horizontal bottom welds in y direction
	XA	0	in	Centroid of arc weld in cutout (= 0 by symmetry)
	π	3.1416		
	YA	11.2732366	in	Centroid of arc weld in cutout in y direction
	ABW	57.349012		Area of side and horizontal welds, sq in
2.897	h		in	Lug weld leg size
	ATOT	168.530722		Sum of weld area, sq in
	ASW	20.48179		Side welds area on inside of cutout, sq in
	ARW	12.8691183		Weld area of arc in cutout, sq in
	AY1	331.804998		Weld areas of outside side welds times centroid, in^3
	AY3	102.40895		Weld areas of inside side welds times centroid, in^3
	AY5	145.076615		Weld area of arc in cutout times centroid, in^3
	AY	1013.50451		Weld area times sum of centroids
	Xbar	0		Group of welds centroid
	Ybar	6.01376709		Group of welds centroid
	R1	9.50273687	in	Outside side welds centroid
	R3	2.24225862	in	Inside side welds centroid
	R5	5.25946948	in	Centroid of arc in cutout group
	S3	1562.13917	psi	Shear stress in weld when vessel is in horizontal position
	JB	2561.78427		Polar moment of inertia of outside side and bottom welds, in^4
	JS	170.681583		Polar moment of inertia of inside side weld of cutout, in^4
	JR	51.4764732		Polar moment of inertia of arc in cutout, in^4
	J	16435.8094		Polar moment of inertia of group of welds, in^4
	M2	7894428.84		Moment in weld, in-lbf
36	H		in	Vertical distance from bottom horizontal weld to lug hole center
	τy	5763.82602	psi	Weld Torsional stress
	C1	11.9862329	in	Extreme fibers above centroid
	C2	6.01376709	in	Extreme fibers below centroid
	C	11.9862329	in	Greater of C1 or C2
	τxx	5757.21343	psi	Lateral torsional stress in weld
	τw	7319.3526	psi	Total lateral shear stress in weld
				VESSEL IN VERTICAL POSITION
	S1	2670.1363	psi	Stress in weld
	$\sigma 1all$	15,200	psi	Allowable weld stress
38,000	σyw		psi	Specified minimum yield stress in weld
	Okay1	2670.1363		If this space is filled then weld stress is acceptable in vertical position
	NOTOK1			If this space is filled then weld stress is NOT acceptable in vertical position
				VESSEL IN HORIZONTAL POSITION
	S2	9316.36264	psi	Stress in weld
	Okay2	9316.36264		If this space is filled then weld stress is acceptable in horizontal position
	NOTOK2			If this space is filled then weld stress is NOT acceptable in horizontal position

FIGURE 8.5(b) The variables sheet for the weld assessment.

A FEW WORDS ABOUT REINFORCING PADS AND
LIFTING LUGS

Lifting lugs are connected to the shell with fillet welds. If lifting lugs are welded to a reinforcing pad, and the pad is connected to the shell with fillet welds, then the capacity of the lifting lugs is a function of the strength of the pad fillet welds, as well as the fillet welds attaching the lug to the pad. Quite often this is not acceptable. Fillet welds can, and will, shear off if the loads exceed their capacity. For lifting lugs, quite often the reinforcing pad is rectangular or square. If the thickness of the shell is considered too small for the capacity of lifting lugs, then two stiffening rings may be welded around the circumference of the shell with the lifting lugs welded between the rings. In this manner the lifting load is distributed around the shell and not localized.

Example 8.3 Evaluating Welds for Top Flange Lifting Lugs

Top flange lifting lugs are mounted on cover plates that are bolted to a nozzle on the centerline of a vessel. They commonly consist of a lug plate welded to a cover plate. One such lug is shown in Figure 8.6(a). As you can see in the figure, the weld sizes vary; hence, the "line" weld concept cannot be used because that method depends on all welds being the same size. The algorithm used for assessing the welds in Figure 8.6(a) is shown in Figure 8.6(b). The corresponding variable sheet is shown in Figure 8.6(c).

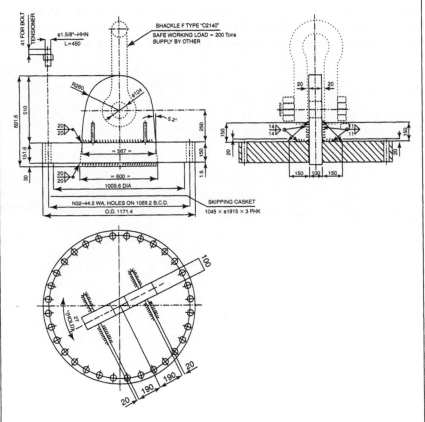

FIGURE 8.6(a)   A top flange lug used to lift a reactor. The lift was successful.

LTOP20mm $= 2 \cdot 567 + 2 \cdot 100$

LTOP14mm $= 4 \cdot 2 \cdot 150 + 4 \cdot 20 + 8 \cdot 150$

LBOT20mm $= 2 \cdot 600 = 4 \cdot 20$

T20mm $= 20 \cdot \text{cosd} (45)$

T14mm $= 14 \cdot \text{cosd} (45)$

WA $=$ (LTOP20mm $+$ LBOT20mm) $\cdot$ T20mm $+$ LTOP14mm $\cdot$ T14mm

$$WC = \frac{(WA) \cdot \sigma_y}{1.35}$$

$$\tau act = \frac{LHmax \cdot 9.807}{WA}$$

$\sigma all = 0.4 \cdot \sigma_y$

If $\tau act < \sigma all$ then $\tau act =$ OK1

If $\tau act = \sigma all$ then $\tau act =$ OK1

If $\tau act > \sigma all$ then $\tau act =$ NOTOK1

AreaT1 $=$ T20mm $\cdot$ 100

AreaS1 $= 2 \cdot 567 \cdot$ T20mm

AreaB1 $=$ T20mm $\cdot$ 100

SUMA1 $=$ AreaT1 $+$ AreaS1 $+$ AreaB1

yT1 $= 283.5$

yS1 $= 0$

yB1 $= -283.5$

yT1SQ $=$ yT1 $\cdot$ yT1

yS1SQ $= 0$

yB1SQ $=$ yB1 $\cdot$ yB1

AyT1 $=$ AreaT1 $\cdot$ yT1

AyS1 $=$ AreaS1 $\cdot$ yS1

AyB1 $=$ AreaB1 $\cdot$ yB1

SUMAy1 $=$ AyT1 $+$ AyS1 $+$ AyB1

AyT1sq $=$ AreaT1 $\cdot$ yT1 $\cdot$ yT1

AyS1sq $=$ AreaS1 $\cdot$ yS1 $\cdot$ yS1

AyB1sq $=$ AreaB1 $\cdot$ yB1 $\cdot$ yB1

SAysq1 $=$ AyT1sq $+$ AyS1sq $+$ AyB1sq

IT1 $= 0$

$$IS1 = \frac{2 \cdot T20mm \cdot \left[567^3 \right]}{12}$$

IB1 $= 0$

I1 $=$ SAysq1 $+$ IT1 $+$ IS1 $+$ IB1

$$Z1 = \frac{I1}{\dfrac{567}{2}}$$

AreaTL $=$ T14mm $\cdot$ 150

AreaSL $=$ T14mm $\cdot$ 20

AreaBL $=$ T14mm $\cdot$ 150

FIGURE 8.6(b)   The equations sheet of the algorithm.

$yTL = 220.0$

$ySL = 230.0$

$yBL = 240.0$

$yTLsq = yTL \cdot yTL$

$yS1sq = ySL \cdot ySL$

$yBLsq = yBL \cdot yBL$

$AyTL = AreaTL \cdot yTL$

$AySL = AreaSL \cdot ySL$

$AyBL = AreaBL \cdot yBL$

$AyTLsq = AreaTL \cdot yTL \cdot yTL$

$AySLsq = AreaSL \cdot ySL \cdot ySL$

$AyBLsq = AreaBL \cdot yBL \cdot yBL$

$SAysq = AyTLsq + AySLsq + AyBLsq$

$ITL = 0$

$$ISL = \frac{T14mm \cdot \left[20^3 \right]}{12}$$

$IBL = 0$

$IL = ITL + ISL + IBL + SAysq$

$IL8 = 8 \cdot IL$

$$ZL = \frac{IL8}{230}$$

$AreaTB1 = T20mm \cdot 100$

$AreaSS1 = 2 \cdot T20mm \cdot 600$

$AreaBB1 = T20mm \cdot 100$

$yTB1 = 300$

$ySB1 = 0$

$yBB1 = -300$

$yTB1sq = yTB1 \cdot yTB1$

$ySB1sq = ySB1 \cdot ySB1$

$yBB1sq = yBB1 \cdot yBB1$

$AyTB = AreaTB1 \cdot yTB1$

$AySS = AreaSS1 \cdot ySB1$

$AyBB = AreaBB1 \cdot yBB1$

$SAyB = AyTB + AySS + AyBB$

$AyTBsq = AreaTB1 \cdot yTB1 \cdot yTB1$

$AySSsq = AreaSS1 \cdot ySB1 \cdot ySB1$

$AyBBsq = AreaBB1 \cdot yBB1 \cdot yBB1$

$SAyBB = AyTBsq + AySSsq + AyBBsq$

$IBT = 0$

FIGURE 8.6(b)   (continued)

$$ISS = \left[\frac{1}{12}\right] \cdot T20mm \cdot 600^3$$

$$IBOT = SAyBB + ISS$$

$$ZBOT = \frac{IBOT}{300}$$

$$ZTOT = Z1 + ZL + ZBOT$$

$$ITOT = I1 + IL + IBOT$$

$$b = e + \frac{tcp}{2}$$

$$PVmax = LVmax \cdot 9.807$$

$$\sigma b = \frac{PVmax \cdot b}{ZTOT}$$

If $\sigma b < 0.6 \cdot \sigma y$ then $\sigma b = OK2$

If $\sigma b = 0.6 \cdot \sigma y$ then $\sigma b = OK2$

If $\sigma b > 0.6 \cdot \sigma y$ then $\sigma b = NOTOK2$

$$\sigma EFF = \beta \cdot \sqrt{\sigma b^2 + 3 \cdot \tau act^2}$$

$$\sigma aEFF = 0.9 \cdot \sigma y$$

If $\sigma EFF < \sigma aEFF$ then $\sigma EFF = OK3$

If $\sigma EFF = \sigma aEFF$ then $\sigma EFF = OK3$

If $\sigma EFF > \sigma aEFF$ then $\sigma EFF = NOTOK3$

FIGURE 8.6(b) (continued)

Input	Name	Output	Unit	Comment
				COMPUTING SHEAR CAPACITY
	LTOP20mm	1334	mm	Length of 20 mm fillet welds on top of cover plate
	LTOP14mm	2480	mm	Length of 14 mm fillet welds on top cover plate
	LBOT20mm	1280	mm	Length of 20 mm fillet welds on bottom
	T20mm	14.1421356	mm	Throat of 20 mm fillet weld
	T14mm	9.89949494	mm	Throat of 14 mm fillet weld
	WA	61518.29		Total weld area, sq mm
	WC	11301137.7	MPa	Weld capacity of welds on top lug
248	σy		MPa	Yield strength of weld metal
	τact	44.1519523		Actual shear stress in welds
276960.6	LHmax		N	Maximum vertical force
	σall	99.2		Allowable shear stress for fillet welds
	OK1	44.1519523		If this space has a value, the τact is acceptable
	NOTOK1			If this space has a value, the τact is not acceptable
				COMPUTING THE MOMENT OF INERTIA IN THE LUG WELDS
				FOR SECTION 1
	AreaT1	1414.21356		Weld area for 100 mm top side, sq mm
	AreaS1	16037.1818		Weld area for 567 mm sides, sq mm
	AreaB1	1414.21356		Weld area for 100 mm bottom side, sq mm
	SUMA1	18865.6089		Sum of 20 mm weld area on top of cover plate, sq mm
	yT1	283.5		Distance from centroid of top weld to lug centerline
	yS1	0		Distance from centroid of side welds to lug centerline
	yB1	−283.5		Distance from centroid of bottom weld to lug centerline
	yT1SQ	80372.25		
	yS1SQ	0		
	yB1SQ	80372.25		
	AyT1	400929.545		
	AyS1	0		
	AyB1	-400929.54		
	SUMAy1	0		
	AyT1sq	113663526		
	AyS1sq	0		
	AyB1sq	113663526		
	SAysq1	227327052		
	IT1	0		
	IS1	429648128		
	IB1	0		
	I1	656975180		Total moment of inertia of welds on top of cover plate, mm^4
	Z1	2317372.77		Section modulus of top center lug plate, cu mm
				FOR THE FOUR LEGS ATTACHED TO SECTION 1
	AreaTL	1484.92424		Weld area on top side of leg, sq mm
	AreaSL	197.989899		Weld area on side of leg, sq mm
	AreaBL	1484.92424		Weld area on bottom side of leg, sq mm
	yTL	220	mm	Distance from centroid of weld to lug centerline
	ySL	230	mm	Distance from centroid of weld to lug centerline
	yBL	240	mm	Distance from centroid of weld to lug centerline
	yTLsq	48400		
	yS1sq	52900		
	yBLsq	57600		
	AyTL	326683.333		
	AySL	45537.6767		
	AyBL	356381.818		
	AyTLsq	71870333.2		
	AySLsq	10473665.6		
	AyBLsq	85531636.3		
	SAysq	167875635		
	ITL	0		
	ISL	6599.66329		
	IBL	0		
	IL	167882235		
	IL8	1343057878		Moment of inertia for welds on top legs, mm^4
	ZL	5839382.08		Section of 8 welded sides of braces, cu mm

FIGURE 8.6(c)   The variable sheet of the algorithm.

Input	Name	Output	Unit	Comment
				FOR WELDS ON BOTTOM OF COVER PLATE
	AreaTB1	1414.21356		
	AreaSS1	16970.5627		
	AreaBB1	1414.21356		
	yTB1	300		
	ySB1	0		
	yBB1	-300		
	yTB1sq	90000		
	ySB1sq	0		
	yBB1sq	90000		
	AyTB	424264.069		
	AySS	0		
	AyBB	-424264.07		
	SAyB	0		
	AyTBsq	127279221		
	AySSsq	0		
	AyBBsq	127279221		
	IBT	0		
	ISS	254558441		
	IBOT	509116882		Total moment of inertia of bottom fillet welds, mm^4
	SAyBB	254558441		
	ZBOT	1697056.27		Section modulus of welds on bottom center plate, cu mm
				COMPUTING THE BENDING STRESS IN LUG WELDS
250	e		mm	Distance from lug hole to top of cover plate
	b	325	mm	Moment arm from hole in lug to center of cover plate thickness
150	tcp		mm	Thickness of cover plate
	ITOT	1333974297		Total moment of inertial for lug fillet welds, mm^4
	ZTOT	9853811.12		
	σb	46.183714	MPa	Actual bending stress in welds
142782.03	LVmax		Kg	Maximum horizontal weight for bending at 10 deg from TK run for top lug 9 Aug 2006
	PVmax	1400263.37	N	Maximum horizontal force for bending
	OK2	46.183714		If this space is filled then σb is acceptable
	NOTOK2			If this space is filled then σb is not acceptable
				TOTAL EFFECTIVE STRESS IN WELDS
	σEFF	80.4034035	MPa	Total stress in welds
.9	β			Factor of yield stress
	σaEFF	223.2	MPa	Allowable stress for effective stress
	OK3	80.4034035		If this space has a value then σEFF is acceptable
	NOTOK3			If this space has a value then σEFF is not acceptable

FIGURE 8.6(c)   (continued)

Example 8.4 Capacities of Various Welds

Often you may be required to evaluate the capacity of different types of welds. Figure 8.7 shows a cross-section of a top flange lug forged to an inner circular plate that is welded to an outer forged cover plate. The originator wanted to use all fillet welds, but a combination of fillet and groove welds produced a much stronger welded piece.

You evaluate the welds as follows:

$$SL = \text{SHEAR RESISTANCE LENGTH}$$

$$SL = 50 + 38 + 50\cos(45°) + 38\cos(45°) = 150.23 \text{ mm}$$

The length resisting the shear is the full length of the groove welds and the weld leg of the fillet weld times cos(45°), which is the fillet weld throat. The fillet weld throat is the length that determines the weld capacity. The view of the cover plate attached to the lug is axisymmetric about the center axis of the configuration. Thus, the weld capacity is as follows:

$$WC = \text{WELD CAPACITY}$$

With a specified minimum yield strength of 248 MPa and an impact factor for lifting of 1.35, you have

$$WC = \frac{\pi D(SL)\sigma_y}{1.35} = \pi(800)(150.23)\text{mm}^2 \left(\frac{248}{1.35}\right)\frac{N}{mm^2}$$

or, $WC = 69{,}360{,}855.2\,N$

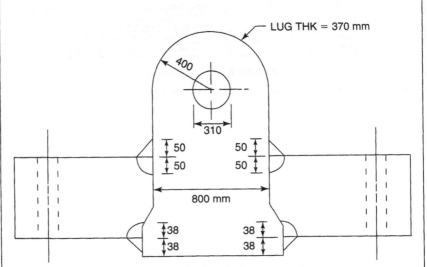

FIGURE 8.7 Capacities of fillet and groove welds.

From prior calculations, you know that the actual load on the welds is 13,243,500 N. The actual shear stress is as follows:

$$\tau = \frac{13,243,500\,\text{N}}{\pi(800)\text{mm}(150.23)\text{mm}} = 35.08\,\text{MPa}$$

The allowable shear stress = $0.40\sigma_y = 99.20\,\text{MPa} > 35.08\,\text{MPa}$.
The maximum weight that the lug can lift is as follows:

$$\text{MAX ALLOW LIFT WEIGHT} = \frac{13,243,500\,\text{N}}{(1.35)(9.807)\dfrac{\text{N}}{\text{Kg}}} = 1,000,305.9\ \text{Kg}$$

The reactor was successfully lifted with the lug. The groove welds added significant strength to the welded configuration.

Chapter 9

Rigging Devices

This chapter provides an overview of various rigging devices used in the field. The plant engineer's responsibilities include the lifting lugs, tail lugs, and trunnions. We have discussed these items at length in previous chapters. Responsibility transfers to riggers when the time comes to lift. It does not hurt for the plant personnel to know some rigging terminology to facilitate communication. During the process of designing lift systems, it is advisable to have riggers informed as to the design. A typical topic is the hole in the lifting lug. Nothing irritates a rigger more than when an engineer specifies that the shackle pin be 1/32" in diameter smaller than the lug hole. With such close tolerance, the pin cannot be removed after the lift. Consequently, construction personnel cut larger holes in the lug to fit the pins—often resulting in uneven holes and high discontinuity stresses in the lug plate.

BLOCKS

A block is a frame that encloses one or several sheaves and has a hook that allows attachment to a vessel or stack or other cargo to a fixed anchor point. The block has two functions: (1) it is used to change the direction of a wire

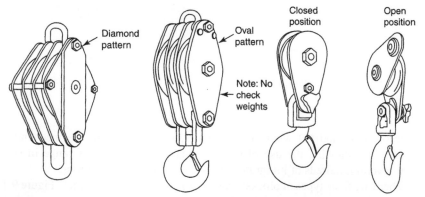

FIGURE 9.1 Different kinds of snatch blocks and wire rope blocks (courtesy of the Bechtel Corporation).

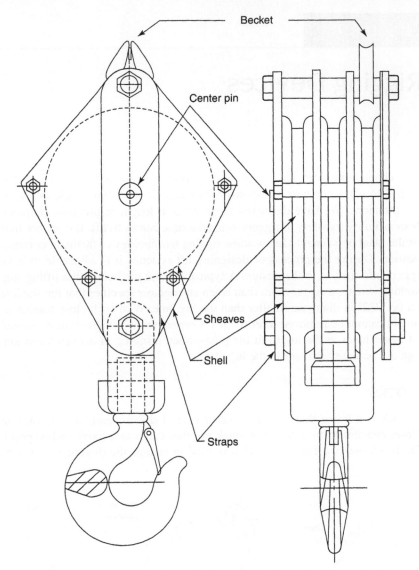

FIGURE 9.2   A typical wire rope block (courtesy of the Bechtel Corporation).

cable or rope; and (2) when used in pairs, blocks increase the mechanical capacity by allowing the use of multiple parts of line. Blocks range in size from several hundreds of kilograms (or pounds) to hundreds of tons.

There are three types of blocks: snatch, wire rope, and crane block. Figure 9.1 shows variations of the snatch block and wire rope block. A typical wire rope block is shown in Figure 9.2.

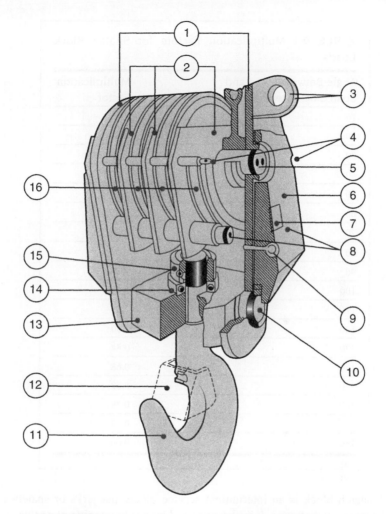

1. Side plates
2. Center plates
3. "Mouse ear" deadend
4. Upper tie bolts
5. Center pin
6. Cheek weight
7. Safety precautions
   plate:
   Tonnage rating
   nameplate
   (opposite sides)
8. Lower tie bolts
9. Cheek weight cap
   screw(s) (1 or 2)
10. Trunnion pin, or
    hook housing
    trunnion
11. Hook
12. Hook latch
13. Hook housing
14. Thrust bearing
15. Hook nut
16. Wire rope sheaves

FIGURE 9.3   A typical crane block (courtesy of the Bechtel Corporation).

TABLE 9.1 Multiplication Factors for Snatch Block Loads

Angle Between Lead and Load Lines (Degrees)	Multiplication Factor
10	1.99
20	1.97
30	1.93
40	1.87
50	1.81
60	1.73
70	1.64
80	1.53
90	1.41
100	1.29
110	1.15
120	1.00
130	0.84
140	0.68
150	0.52
160	0.35
170	0.17
180	0.00

A snatch block is an intermittent service block that jerks or snatches the load over small distances. It is characterized by a side-opening plate that facilitates threading wire rope through the block.

A crane block, as opposed to a snatch block, is required to perform long lifts under continuous service conditions. A crane block has multiple large diameter sheaves, designed for long service life, with check plate weights added to the block side frame to increase the overhaul weight. Normally, a crane block is outfitted with swivel hooks that allow the mass being lifted to be rotated without fouling the multiple parts of reeving. Figure 9.3 shows a typical crane block.

SELECTION OF A BLOCK

The governing criterion for selecting a block is the load to be encountered rather than the diameter or strength of the rope used. In blocks with multiple sheaves, the load is distributed among several parts of the rope, whereas the

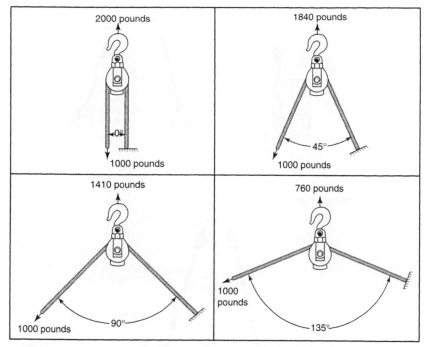

FIGURE 9.4   Rope angle as a variable of snatch block loads (courtesy of the Bechtel Corporation).

shackles or hooks on the blocks must carry the entire load. For heavy loads and fast hoisting, it is recommended that roller or bronze bearings be used. The block anchor supports the total weight of the load, plus the weight of the blocks and load applied to the lead line.

Snatch blocks are either single or double sheave blocks manufactured with a shackle eye, hook, and swivel end fittings. Snatch blocks are mostly applied for altering the direction of the pull on a line. The stress on a snatch block varies with the angle between the lead and load line. When the two lines are parallel, 2,000 lbs on the lead line added to the 1,000 lbs on each load line result in a load of 4,000 lbs on the block. Table 9.1 shows the multiplication factors for snatch block loads. Figure 9.4 illustrates the rope angle as a variable of snatch block loads.

LIFTING AND ERECTING PRESSURE VESSELS AND STACKS

Figure 9.5 illustrates a typical method of lifting and erecting pressure vessels and stacks where lifting lugs are attached.

The spreader bar shown in Figure 9.5 avoids any bending moment on the lift lugs because the choker angle is 90°. As mentioned in Chapter 7, the choker angle should never be more than 45° under any circumstances. It is

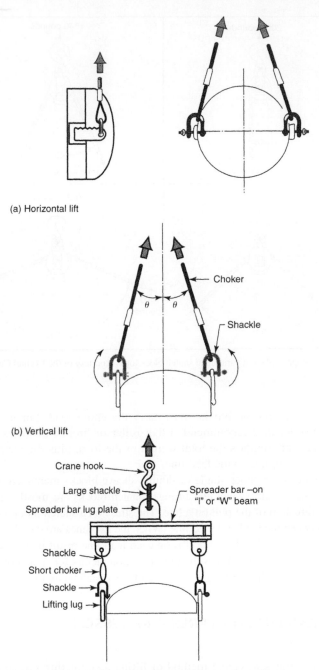

(a) Horizontal lift

(b) Vertical lift

(c) Spreader bar rig avoids excessive bending moments on lifting lugs

FIGURE 9.5   Lifting lugs and erection procedure.

highly recommended that the choker angle be as small as possible—normally 15°. Many industrial standards use 30° as the maximum choker angle.

SHACKLES

Shackles are the most common devices used with lifting lugs. There are three basic types: wide body, bolt type, and screw pin shackles. Figure 9.6 shows alloy wide body shackles.

"WIDE BODY" SLING SAVER SHACKLES INCREASE SLING LIFE

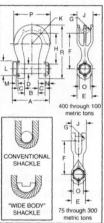

- Greatly improves wearability of wire rope slings.
- Can be used to connect HIGH STRENGTH Synthetic Web Slings, HIGH STRENGTH Synthetic Round Slings or Wire Rope Slings.
- Increase in shackle bow radius provides minimum 58% gain in sling bearing surface and eliminates need for a thimble.
- Increases usable sling strength minimum of 15%.
- Pin is non-rotating, with weld on handles for easier use (300 ton and larger).
- All ratings are in metric tons, embossed on side of bow.
- Forged alloy steel from 30 through 300 metric tons.
- Cast alloy steel from 400 through 1000 metric tons.
- Sizes 400 tons and larger are tested to 1.33 times Working Load Limit.
- Sizes 300 tons and smaller are proof tested to 2 times the Working Load Limit.

G-2160

- All 2160 shackles are individually proof tested, Crosby certification available at time of order. Shackles requiring ABS, DNV, Lloyds and other certifications are available upon special request and must be specified at time of order.

Patented

- Shackles are produced in accordance with certified lifting appliance requirements.
  - Non Destructive Testing
  - Serialization/Identification
  - Material Testing (Physical/Chemical/Charpy)
  - Proof Testing
- All sizes Quenched and Tempered for maximum strength.

- Bows and pins are furnished Dimetcoted. All Pins are Dimetcoted then painted red.
- Type Approval and certification in accordance with DNV specifications 2.7-1 Offshore Containers and DNV rules for Lifting Appliances-Loose Gear.

NOTICE: All G-2160 shackles are magnetic particle inspected

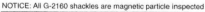

Working Load Limit* (metric tons)	G-2160 Stock No.	Weight Each (lbs.)	Dimensions (Ins.)										
			A	B ± .25	C	D ± .02	E	G	H	J	K	P	R
† 30	1021575	25	7.75	2.38	1.38	1.63	3.56	2.00	6.50	3.13	2.50	9.13	11.00
† 40	1021584	35	9.06	2.88	1.75	2.00	4.00	2.31	8.06	3.75	3.00	10.62	13.62
† 55	1021593	71	10.41	3.25	2.00	2.26	4.63	2.63	9.38	4.50	3.50	12.88	15.53
† 75	1021290	99	13.62	4.13	2.12	2.76	4.76	2.52	11.41	4.72	3.66	12.32	18.31
† 125	1021307	161	15.75	5.12	2.56	3.15	5.71	3.15	14.37	5.90	4.33	14.96	22.68
† 200	1021316	500	20.00	5.90	3.35	4.13	7.28	4.33	18.90	8.07	5.41	19.49	29.82
† 300	1021325	811	23.27	7.28	4.00	5.25	9.25	5.51	23.62	10.43	6.31	23.64	37.39
†† 400	1021334	1041	28.13	8.66	5.16	6.30	11.02	6.30	22.64	12.60	7.28	27.16	38.78
†† 500	1021343	1378	31.87	9.84	5.59	7.09	12.52	6.69	24.80	13.39	8.86	31.10	42.72
†† 600	1021352	1833	35.94	10.83	6.04	7.87	13.78	7.28	27.56	14.57	9.74	34.06	47.24
†† 700	1021361	2446	39.07	11.81	6.59	8.46	14.80	7.87	28.94	15.75	10.63	37.01	50.18
†† 800	1021254	3016	38.82	12.80	7.19	9.06	15.75	8.27	29.53	16.54	10.92	38.39	52.09
†† 900	1021389	3436	41.34	13.78	7.78	9.84	16.93	8.66	29.80	17.32	11.52	40.35	54.04
††1000	1021370	4022	46.30	14.96	8.33	10.63	17.72	9.06	29.92	18.11	12.11	42.32	55.3

● Ultimate is 5 times the Working Load Limit.
† Forged Alloy Steel. Proof Load is 2 times the Working Load Limit.
†† Cast Alloy Steel. Proof Load is 1.33 times the Working Load Limit.

FIGURE 9.6 Typical catalog of wide body shackles (courtesy of Slingmax).

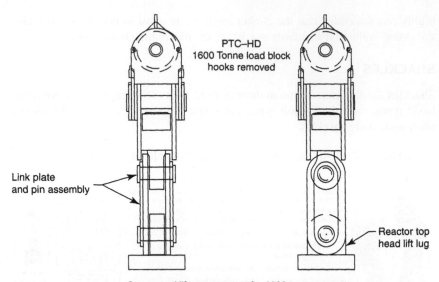

Conceptual lift arrangement for 1000 tonne+ reactors

FIGURE 9.7   Link plate and pin assembly arrangement.

In cases where very high loads are encountered, such as the flange lug mounted on a cover plate bolted to a nozzle on top of the vessel, a link plate and pin assembly is often used. This arrangement is shown in Figure 9.7.

The advantage of this device over a shackle is that it can fit in tighter spaces; sometimes the shackle can interfere with the cover plate bolts. It can be designed and fabricated for loads beyond conventional commercially available shackles. The arrangement in Figure 9.7 is designed to lift 1600 tonnes (or metric tons).

Other rigging devices, such as hooks, various cables, and types of cranes, can be found in the following recommended sources:

1. *Slingmax Rigging Handbook,* by II Sling, Inc.
2. Products of Industrial Training International, Inc.
   - *Crosby User's Lifting Guide*
   - *Mobile Crane Operator Reference Card*
   - *Equipment Operator's Card*
   - *Rigging Gear Inspection Card*
   - *Journeyman Rigger's Reference Card*
   - *Master Rigger's Reference Card*
   - *Lineman Rigger's Reference Card*
3. *Yellow Strand Wire Rope Handbook,* by Broderick & Bascom Rope Company, 10440 Trenton Ave St. Louis, Mo 63132

Index

2:1 ellipsoidal head weights, 25–6
Air distributor piping, 92
American Engineering System
    (AES), 1
American National Standard
    Institute (ANSI) safety
    codes, 117–18
American Society of Mechanical
    Engineers (ASME), 1, 4
  flanged and dished heads, 26–7
ANSI standards, *see* American
    National Standard Institute
    (ANSI) safety codes
ASCE 7-2005, 75
ASME F&D head, 26–7
ASME STS-1, 4, 45, 46, 47, 54, 55,
    64, 65, 74
ASTM SA-333 pipe material, 67

Basic wind speed, 79
Bearing failure, 119, 129
  validation tests for, 130–1
Bellows expansion joints, 109
Bending moments and torque, 5–6
Blocks, 181–4
  selection, 184–5
Bridles and center of gravity, 116
British Thermal Unit (BTU), 8

Capacity reduction factor, 122, 123
Charpy impact test, 7
Circumferential joints, *see*
    Longitudinal stress
Circumferential stress:
  conical sections, 17–18
  in cylindrical shell, 12–13
Clam shell, 109

Coefficient of thermal expansion, 8
Combined stress, 99
Conical sections:
  circumferential stress, 17–18
  conical transitions, 20
    flair section, 21
    half-apex angle, computation
      of, 24
    knuckle section, 19, 21
    pressure-area force balance
      procedure, equations based
      on, 21–2
  final values, 19
  knuckle section, 19
  longitudinal stress, 18
Correlation length, 41
Crane block, 182, 183, 184
Critical buckling stress, 128
Critical lift, 116
Critical wind velocity, 42, 65
Cylindrical shells, 11
  circumferential stress, 12–13
  final/resulting values, 14
  longitudinal stress, 13–14

Damping pads, 62–4
Density, 5
Double plane shear failure, 119,
    125–7
Dynamic response, of pressure
    vessels and stacks, 41
  flow-induced vibration-impeding
    devices:
    damping pads, 62–4
    helical strakes, 56–62
    ovaling rings, 64–6
  guy cables, 71–2

Dynamic response, of pressure
    vessels and stacks (*Contd.*)
    basic methodology, 66–71
    screening criteria, 42–56

Effective wind diameter, 82–4
Elliptical head, 14–15
    partial volumes of, 33–6
Energy, units of, 7
    thermal conductivity units, 8
    thermal expansion, coefficient
        of, 8
Expansion joints, *see* Bellows
        expansion joints

Fabreeka International, 62
Fillet and groove welds, capacities
        of, 178–9
Finite element method (FEM), 123
Flare stacks, 66, 67, 68
    see also Guy cables
Flexible structure, 79–80
Flow-induced vibration-impeding
        devices:
    damping pads, 62–4
    helical strakes, 56–62
    ovaling rings, 64–6
Fluid-structure interaction, 41
    criteria, 47–56
Force, 2
Force, pound, and second system
        (FPS) system, 3
Forces on internal components, 93–7
Formulas for pressure vessels, 11
    ASME F&D head weights, 26–7
    conical sections:
        circumferential stress, 17–18
        conical transitions, 19–22, 24
        final values, 19
        longitudinal stress, 18
    cylindrical shells, 11
        circumferential stress, 12–13
        final or resulting values, 14
        longitudinal stress, 13–14

elliptical head, 14–15
head weight computations, 25–6
hemispherical head weights, 27–8
partial volumes, 28
    of cylinder in horizontal
        position, 29
    of elliptical heads, 33–6
    of hemispherical head, 30–1
    of spherically dished heads,
        31–3
    torispherical heads, 36–40
spherical shell, 14
torispherical head, 15–17

Groove and fillet welds, capacities
        of,
        178–9
Gross ton, *see* Long ton
Gust-effect factor, 79–82
Guy cables, 66–72

Half-apex angle, computation of, 24
Head weights, formulas for:
    2:1ellipsoidal head weights, 25–6
    ASME F&D head weights, 26–7
    hemispherical head weights, 27–8
Heavy lift, 116
Helical strakes, 56–62
Hemispherical head, 14
    partial volume of, 30–1
Hemispherical head weights, 27–8
Hoop tension, 119, 124–5

Impact factor, 120, 166, 170, 178
Importance factor, 79
Internal assessment, of pressure
        vessels, 91
    forces on internal components,
        93–7
    internal expansion joints, 109
    lined plates and internal
        components, 97
    structural formulations:
        support clips, 99–104

tray support ring, 98
wire and sheet metal gauges,
        105–8

Joule, 5–6, 7

KiloNewtons (KN), 4–5

Lift categories, 116
Lifting and erecting pressure vessels
        and stacks, 185–7
Lifting and tailing devices:
    failure modes, 119, 121
        bearing failure, 121, 129–31
        double plane shear failure, 119,
            121, 125–7
        evaluation, 132–6
        hoop tension, 119, 121, 124–5
        out-of-plane instability
            (dishing) failure, 119, 127–9
        tension at net section, 119, 120,
            121, 122–4
    multiple loads, 137–42
    pin hole doubler plates, 131–2
    rigging analysis, 143–50
    trunnions, 150–62
    see also Lifting lugs
Lifting lugs, 163
    and erection procedure, 186
    and reinforcing pads, 171
    with simple "U" design, 164
        equations sheet for weld
            assessment, 165
        variable sheet for weld
            assessment, 166
    top flange lifting lugs, 172–7
    with "U" shape and cutout, 167
        equation sheet for weld
            assessment, 168–9
        variables sheet for weld
            assessment, 170
    see also Lifting and tailing
        devices
Light lift, 116

Lined plates and internal
        components, 97
Long ton, 111, 112
Longitudinal joints, see
        Circumferential stress, in
        cylindrical shell
Longitudinal stress, 13–14, 18

Mass, 2, 3
Mass damping parameter, 45–6,
        54–5
Maximum allowable pressure
        (MAP), 11
Maximum allowable working
        pressure (MAWP), 11
Medium lift, 116
Meter, kilogram, and second (MKS)
        system, 2
Metric SI system, 1, 2
    in Australia, 114
    bending moments and torque,
        5–6
    density, 5
    familiarity, 4–5
    ton, concept of, 111
    warning about combining, 6–7
Metric ton, 111–12
MTBR Regenerator air distributor,
        internal collapse of, 91–3
Multiple loads, on lifting and tail
        lugs,
        137–42
    rigging analysis, 143–50

Natural frequency method, 43–5,
        49–54, 65
Net ton, see Short ton

Occupational Safety and Health
        Administration (OSHA),
        117
Out-of-plane instability (dishing)
        failure, 119, 127–9
Ovaling rings, 64–6

Partial volumes pressure vessel
    calculations, 28
  of cylinder in horizontal position,
    29
  of elliptical heads, 33–6
  of hemispherical head, 30–1
  of spherically dished heads, 31–3
  torispherical heads, 36–40
Pin hole doubler plates, 131–2
Pound, 3, 4
Pressure, unit for, 6
Propagative compression, 95–7

Rayleigh's method, 43, 65
Reinforcing pads and lifting lugs,
    171
Remediation devices, 66
Rigging analysis, of pressure vessel
    lifting, 143–50
Rigging devices, 181
  blocks, 181–4
    selection, 184–5
  lifting and erecting pressure
    vessels and stacks, 185–7
  shackles, 187–8
Rigid structure:
  guest-effect factor, 79

Safety considerations, for lifting and
    rigging, 111
  American National Standard
    Institute safety codes,
    117–18
  bridles and center of gravity, 116
  lift categories, 116
  preparation for lift, 116–17
  slings, maximum capacity of,
    112–15
  weight ton, concept of, 111–12
Saudi Arabia, 66, 70, 72, 73, 116
Screening criteria, of dynamic
    resonance response, 42–56
Shackles, 120, 185, 187–8
Shamal, 79

Short ton, 111, 112
Slings, maximum capacity of,
    112–13
Slow compression, see Spatially
    uniform compression
Slug, 3, 4
Snatch block, 181, 184, 185
Spatially uniform compression, 95,
    96
Spherical shell, see Hemispherical
    head
Spherically dished head:
  partial volumes of, 31–3
Stack helical strake design, 60–1
Stress, unit for, 6
Structural damping coefficient,
    43–45
Structural formulations:
  support clips, 99–104
  tray support ring, 98
Support clips:
  with applied tensile force, 100–1
  with continuous fillet weld with
    in-plane bending moment,
    102, 104
  with continuous fillet weld
    with out-of-plane bending
    moment, 102
  with in-plane bending moment,
    101, 103
  with out-of-plane bending
    moment, 101, 102
  welded on two sides with fillet
    welds, 99, 100
Surface roughness, 79
Systems of units, 1
  energy, units of, 7
    coefficient of thermal
      expansion, 8
    thermal conductivity units, 8
  metric SI system, 1, 2, 3
    basic units, 5
    bending moments and torque,
      5–6

density, 5
familiarity, 4–5
warning about combining, 6–7
stress and pressure, units for, 6
toughness, unit of, 9–10

Tech Products Corporation (TRC), 62
Tensile force, 69, 71
Tensile splitting, *see* Hoop tension
Tension at net section, 119, 120,
    122–4
Thermal conductivity units, 8
Thermal expansion:
    coefficient, 8
    of flare stack, 70
Ton, concept of, 111–12
Tonne, *see* Metric ton
Top flange lifting lugs, 172–7
Topographic factor, 78–9
Torispherical heads, 15
    geometrical equations, 16–17
    in horizontal position, 38–40
    partial volumes of, 36–8
Toughness, unit of, 9–10
Tray support ring, 98
Trunnions, 150–62, 171

U.S. Customary Units (USCU), 1

Velocity pressure coefficient, 77–8
Velocity pressure distribution, 76
Vortex shedding frequency, 42

Weight ton, *see* Long ton
Weld attachments, assessing:
    fillet and groove welds, capacities
        of, 178–9
    lifting lugs, 163
        and reinforcing pads, 171
        with simple "U" design,
            164–6
        top flange lifting lugs, 172–7
        with "U" shape and cutout,
            167–70
Weld stresses, 99, 159
Wide body shackles, 187
Wind directionality factor, 76, 85
Wind loading, on pressure vessels
        and stacks, 75, 85–9
    basic wind speed, 79
    gust-effect factor, 79–82
    importance factor, 79
    projected area normal to wind,
        82–4
    topographic factor, 78–9
    velocity pressure coefficient,
        77–8
    velocity pressure distribution, 76
    wind directionality factor, 76
Wire and sheet metal gauges,
        105–8, 109
Wire rope block, 181, 182

Printed and bound by CPI Group (UK) Ltd, Croydon, CR0 4YY

03/10/2024

01040433-0011